THEORETICAL PROBLEMS
OF GEOGRAPHY

THEORETICAL PROBLEMS
OF GEOGRAPHY

V. A. ANUCHIN

Edited by:

Roland J. Fuchs

George J. Demko

Translated by:

Steven Shabad

With an Introduction by:

David Hooson

OHIO STATE UNIVERSITY PRESS: COLUMBUS

Library of Congress Cataloging in Publication Data
Anuchin, V A
    Theoretical problems of geography
    Translation of Teoreticheskie problemy geografii.
    Bibliography: p.
    1. Geography. I. Title.
G70.A713   910'.01   77-8437
ISBN 0-8142-0221-7

CONTENTS

Contents

EDITORS' PREFACE

The publication in 1960 of V. A. Anuchin's *Theoretical Problems of Geography** must be viewed in retrospect as a major event in the history of Soviet geographical thought. Anuchin's work is important in a number of respects that justify making it available to an English-reading audience:

1. It remains the most significant Soviet statement on the theory and philosophy of geography. Anuchin examines the history of the development of geographic thought from the standpoint of Marxism-Leninism and proposes a philosophy and theory for a *unified* geography. In the process he adds interesting new insights and dimensions to works currently available in Western languages.

2. The work has achieved unusual stature as a polemical study. As the most comprehensive and forceful Soviet statement on the need for a unified or monistic geography, it has generated an unparalleled and continuing debate in Soviet

---

*V. A. Anuchin, *Teoreticheskiye problemy geografii* [Theoretical Problems of Geography], Gosudarstvennoye Izdatel'stvo Geograficheskoy Literatury [State Publishing House of Geographical Literature] (Moscow, 1960).

academic and other circles. Ironically the English-language reader has been able to follow much of this debate through translated articles that have appeared in *Soviet Geography; Review and Translation* and other journals, while denied access to the original treatise that is the source of the controversy.

3.   Anuchin's argument in favor of the unity of geography, his stress on the geographical environment as the proper object of geographical study, and the need for greater emphasis on practical and applied work are current and relevant themes in Western geographical circles.

For these reasons then it seems highly appropriate that *Theoretical Problems of Geography* be added to the geographical literature in the English language. Although it is likely to become a basic reference work for student of geography and Marxism-Leninism, it is not without difficulties for the reader who will find it has faults typical of much philosophical and polemical writing — prolonged definitional discourses, exaggeration and overgeneralization, and repetition. The translation that is offered here is essentially a literal one; a freer translation could have conceivably improved the ease of reading but at the expense of distorting the nuances and finer points of the philosophical arguments.

For the convenience of the reader several additions accompany the original Anuchin monograph:

1.   An introduction by Professor David Hooson, a leading scholar of Soviet geographical thought and close observer of the Anuchin controversy, which provides the setting in which the book appeared, and a summary of its reception and significance.

2.   A summary of the major conclusions of the book in Appendix 1; this will permit the reader to review quickly the major points developed at length in the original study.

3.   A selected bibliography of commentaries on Anuchin's work (Appendix 2) that have appeared in the English-language journals, which will permit the interested reader to trace the evolution of the debates generated by Anuchin's work.

The editors, in bringing Anuchin's work to a new audience hope that another significant step will be accomplished in the international exchange of ideas among scholars. If new insights result and Soviet-Western dialogues are enhanced, our goal will have been achieved.

Roland J. Fuchs
George J. Demko

ACKNOWLEDGMENTS

The editors gratefully acknowledge the assistance of the University of Hawaii Foundation, which provided a grant to cover the cost of initial translation, and the secretarial staffs of the geography departments of the Ohio State University and the University of Hawaii. A great debt of gratitude is owed also to Ms. Sally Millett of the Ohio State University Press for her acute editorial skill and patience in handling the manuscript and editors. Finally, we extend our special appreciation to our wives, Gaynelle Ruth and Jeanette Edwina, for serenely enduring the practical problems caused by our involvement with theoretical problems.

PART ONE

INTRODUCTION

INTRODUCTION TO THE ENGLISH EDITION

By David Hooson

Although more than a decade has gone by since Anuchin's book appeared, the intellectual climate and the state of development of world geography seem to make the 1970s, if anything, an even more propitious time for putting out an English translation than prevailed immediately after its original appearance in the Soviet Union.

In the first place, no one can fail to have become conscious of the sudden revelation of "environment" or "ecology" as urgent, if rather vague concepts in the minds of a broad spectrum of the population in the industrialized countries. The fact that the environment is increasingly being apprehended as a pervasively humanized, rather than merely a physical category, greatly broadens the potential significance of Anuchin's elaboration of his central concept — the *geographical environment*. On the plane of academic geographical thought, there can be little doubt that a parallel qualitative change of fundamental importance is taking place in America and certain other countries. In some respects it may be said to be "counterrevolutionary," in relation to the so-called quantitative revolution of a decade or so ago. Rather like the Romantic Protest of the early nineteenth century against the intellectual sway of the Enlightenment, a reaction seems to have set in against what is now seen as an unduly mechanistic preoccupation with technique, precision, analysis of contemporary systems, and over-emphasis on economic at the expense of social, historical, or environmental factors. Alongside this development has been a revival of interest in geographical theory conceived more in terms of long-term purpose and philosophy than short-run method and technique, though in the latter case, one hopes, taking care to preserve and incorporate whatever insights of lasting significance have come out of the quantitative revolution. Thus, Anuchin's book, surveying the theoretical scene — past and present — of geography as a whole, and carefully building up and elaborating his unifying theme — geographical environment — strikes several chords in tune with the developing needs and spirit of the subject in American and other Western countries.

More specifically, the book is an original document of key significance to the recent development of Soviet geography, having stimulated one of the most far-reaching and sharp debates in Soviet academic circles in modern times, comparable with the Lysenko controversy in biology. Moreover, though not always conceded, it seems to have been instrumental to no small degree in changing basic theoretical and practical positions, both in academic geography and the Soviet planning agencies.

Quite apart from these two compelling topical reasons for putting out the book in English at this time, there is the plain fact, which would in itself justify a translation, that the book remains the only one in Soviet history to set out to investigate the theoretical basis of geography as a whole through historical and philosophical analysis, while coming to definite conclusions about what geography is and where it should go from here. This said, it remains to outline briefly the historical context in which the book appeared, and its astonishingly mixed reception in the 1960s.

THE INCUBATION PERIOD

It is necessary to recall the atmosphere in the Soviet Union in the middle 1950s in order to appreciate why and how the book was written. All aspects of Soviet intellectual life in the years following the death of Stalin experienced the release of a ferment of reforming ideas and vigorous argumentation that had lain dormant or muted for a quarter-century. As a result of the Stalinist period, geography had become overwhelmingly physical in character[1] and the nonphysical aspects of the subject were in some danger of being squeezed out and absorbed by other disciplines. Moreover the prevailing doctrine maintained the clear and necessary separation of physical from economic (i.e., all nonphysical) geography, with their supposedly mutually exclusive sets of laws, so that the integrated study of man in relation to his environment, long a central theme in period, was ruled theoretically illegitimate.

The depth of conviction shown by Anuchin and other geographers about the urgency of radical reform and restructuring of geographical theory owed much to their keen awareness of a broken heritage in Russian geography — and the lateness of the hour. For geography had had a long and distinguished history in Russia before the Stalin era[2] and it seemed improbable that any of the founding fathers of the subject at that time would have approved of the turn subsequently taken by Soviet geography, any more than would Lenin, Plekhanov, Engels, or even Marx himself.

Thus a thoroughgoing "thaw" in Soviet geographical theory was taking place in the mid-fifties, as practitioners reestablished spiritual contact with their academic forebears (for the Russians are at least as nationalistic as anyone else!), redefined their own positions, and in general effected a shift in the center of gravity of the subject toward the human side. Anuchin turned out to be the most articulate and uncompromising in staking out new paths for the subject and in reinterpreting the past, though obviously not to the satisfaction of everyone concerned.

These few years preceding the appearance of Anuchin's book in 1960 were, then, fluid — even heady — ones for Soviet geography, during which vigorous new periodicals appeared, the volume of publications grew rapidly, and the content of geography became steadily more balanced. International contacts were eagerly reestablished — 1956, for instance, saw Soviet geographers attending their first international geographical congress for two decades and also reviewing at great length the volume on *American Geography: Inventory and Prospect*. The not unsympathetic tenor of this and other reviews of foreign work contrasted strikingly with the xenophobic vituperation that had been common form a few years before.[3] These early years of the Khrushchev decade, in spite of their inconsistencies and "hare-brained schemes" (as they were afterward dubbed) can now be looked back upon as, in relative terms, something of a Golden Age for the Soviet intellectual. It was a period of confident ebullience, with the great psychological boost of Sputnik as well as destalinization and the publication of books by previously (and subsequently) banned writers, alongside the radical restructuring of the economy and a more outgoing and relaxed foreign policy (before the setbacks of the Sino-Soviet rift and the Cuban missile crisis). Thus 1960 was as good a time as any in recent years to publish a controversial book challenging established academic doctrine in the Soviet Union.

THE BOOK'S IMMEDIATE RECEPTION

The relatively tolerant political atmosphere of the time notwithstanding, it is doubtful whether Anuchin's book would have seen the light of day if it had not possessed one or two influential supporters among the senior geographers, the most important of whom was N. N. Baranskiy. At the age of eighty he was regarded with a unique mixture of respect and affection by most of the profession, and his background and authority were unassailable. An early revolutionary from Siberia and an acquaintance of Lenin, he was at the same time a humane scholar, thoroughly steeped in the geographical traditions of Russia and the West. He retained considerable objectivity and academic sobriety throughout, while dealing, unlike most geographers of his time, with vulnerable human topics. For a generation he was the major figure in economic geography (taken in a broad sense) at Moscow University and, through his long-lived textbooks, in the secondary schools as well. Thus when Baranskiy, in an early review of Anuchin's book, characterized it as "a courageous and . . . a deeply scientific work . . . let us hope that it will be much in demand, both in the Soviet Union and abroad . . . its great value is completely obvious,"[4] the impending battle was bound to be one of titans. For as Yu. G. Saushkin, another influential, but younger, supporter, said, "It required a great deal of boldness to write such a book, since very authoritative scholars, periodicals, and publications have spoken out against a 'unified geography,' dubbing it 'theoretically shameful.' "[5]

Of these authoritative scholars, two have loomed largest in this case, taking the lead in the opposition to Anuchin's main thesis, I. P. Gerasimov and S. V. Kalesnik. Both are physical geographers who had played a major role in the elaboration of the doctrine of the theoretical separation of physical and economic geography. They are the only two geographers who are full members of the Academy of Sciences of the USSR and hold the two most powerful official positions in Soviet geography — director of the Academy's Institute of Geography and president of the All-Union Geographical Society respectively. Both have been strong opponents of Anuchin's, both theoretically, in print in the journals of their own institutions, and practically, as influential members of the councils of the faculties of geography at Moscow and Leningrad universities, to which Anuchin's book was submitted for the Doctorate of Geographical Sciences. His bid for the doctorate at Leningrad University was unanimously turned down in 1961. The following year he submitted it to Moscow, where the public dissertation defense was a dramatic affair attended by hundreds of people. By that time the case had become a *cause célèbre* and rank-and-file sympathy on his home ground was strong. Nevertheless the final vote fell just short of the required two-thirds majority (though later on Anuchin was quietly awarded the doctorate, on a technicality).

Fortunately for English-speaking readers, the monthly journal *Soviet Geography: Review and Translation* (edited by Theodore Shabad and published by the American Geographical Society of New York) was born in 1960, the same year as Anuchin's book. Since one of its chief aims was to introduce Americans and others to the variety of opinions on theoretical matters held by Soviet geographers, it is hardly surprising that this journal has been sprinkled with articles pertaining to the controversy sparked by Anuchin, still continuing today but occurring most thickly in the early 1960s. Anyone who wishes to gauge the intensity of the impact of Anuchin's writings on the whole Soviet geographical profession, and to get some inkling of the *nuances* of the complex issues as well as the line-up of individuals and to some extent, institutions, should make a point of following through these discussions, which are listed in the bibliography at the end of this book. Space forbids anything like a comprehensive analysis of this discussion here, but it is an essential adjunct to the book itself and is concerned with problems, notably the proper understanding of the significance of environment and the difficulties of integrating successfully the heterogeneous phenomena of geography, which are as perennially interesting and important in Western as well as in Soviet geography.

One should have a proper feeling for the deadly seriousness of these discussions, the courage frequently needed, the amount at stake in the challenge to existing doctrine, and the fluctuating, unpredictable character of the political climate within which the arguments had to be carried on. The sensitive Western reader can scarcely fail to detect unfamiliar innuendos that seem to have been hangovers from the fear-ridden Stalinist years. In this controversy, the chief "Achilles' heel" of Anuchin and some of his supporters, which seems to have been persistently aimed at by their opponents, is any presumed association or mutual approbation with "bourgeois," particularly American geographers. One quotation from Kalesnik deserves to be mentioned here, both because of its obvious relevance to our present enterprise and because of its innuendos of the kind that still cast a cloud over Soviet intellectual endeavors. He says "Anuchin's book is bound to be well received abroad. It will undoubtedly be translated into foreign languages because all foreign adherents of a unified geography will seize upon his book as a sensation, especially piquant because it originates in the Soviet Union where, according to general consensus, the tombstone of a unified geography has long been overgrown with weeds."[6] All this, in spite of the fact that Kalesnik was presumably glad, along with the other editors of *Soviet Geography: Accomplishments and Tasks,* to have that volume published in English that same year. Similarly, O. A. Konstantinov, in the course of a particularly virulent review of Anuchin's book, after quoting the latter's relatively sympathetic references to some American geographers says "Now we know to whom V. A. Anuchin appeals and with whom he has something in common."[7] Attempts were even made in the course of his dissertation defense at Moscow University to smear Anuchin by citing passages, supposedly kindred in thought to his, from the author of the present introduction, who had just been dubbed "a reactionary American geographer" by Academician Gerasimov at that same defense. One can find many other examples of guilt by association with foreign geographers in the annals of this controversy, all of them, it seems, perpetrated by opponents of Anuchin. Similarly Saushkin, an early supporter of Anuchin's, was severely pilloried by a group of economists and geographers who, in a collective letter to a Soviet journal, charged him with "distortion" and traitorous statements in an article that he published in an American journal *(Economic Geography)* in 1962. Saushkin, in his hard-hitting reply, deplores "the fact that some geographers still have not rid themselves of the rough methods of unsubstantiated accusations and intimidations used in the period of Stalin's personality cult."[8]

THE ARENA BROADENS

When the argumentation in the geographical journals had gone on apace for two or three years and, together with the publicity surrounding Anuchin's dissertation defenses, had made the issues seem well known as well as relatively intractable in Soviet academic circles, a leading ideological spokesman of the Communist Party stepped in and gave, as is customary in such cases, an *ex-cathedra* statement before the Presidium of the Academy of Sciences.[9] In it L. F. Ilyichev denounced the Stalinist definition of the environment as "a purely natural category" and the fact that this edict seemed to remain the theoretical pretext for the construction of "an insurmountable wall" between nature and society. Although he chided Anuchin for apparently wanting to include all aspects of society itself in the concept of geographical environment, the general thrust of Ilyichev's pronouncement, in the context of the disposition of authority in the USSR, indicated an unmistakable setback for the established theoretical doctrine of "legal separation" of physical and economic geography espoused by Gerasimov, Kalesnik, and others. Encouraged by this apparent turning of the tide in his favor, Anuchin thenceforth extended his range to cover not only other academic periodicals, notably in philosophy, but also various organs of the semipopular national press.

The most wide-ranging of these forays was the series of six articles in *Literaturnaya Gazeta,* which ran from February to June 1965[10] initiated by Anuchin and completed by Saushkin with the explicit support of the editors. Although the articles represented a variety of views and showed sharp disagreements, the overall result of the series was to subject to outright censure the general policies and priorities of the Institute of Geography of the Academy of Sciences and its director, Gerasimov, particularly the overemphasis on physical geography and the lack of provision for integrated regional and economic studies. Gerasimov was severely criticized by some of his own staff at the institute and also by the Presidium of the Academy of Sciences itself, and urged to shift his institute's emphasis sharply towards economic geography. Saushkin in his concluding article, made a proposal, with the blessing of the editors, that economic geographers should logically be "the conductors of the geographical orchestra," guiding the research priorities in the physical branches of the subject as well. Considering that the structures of both the Institute of Geography and the Moscow State University were still heavily and traditionally loaded in favor of the physical specialities, these proposals must have seemed revolutionary. Yet, if one thinks about it, they are so obviously in line with the traditional precepts and spirit of Marxism that the wonder is that such disharmony between philosophy and practice had been allowed to develop in the period of total Soviet planning.

A NEW PARADIGM FOR SOVIET GEOGRAPHY?

On the face of it, the contrasts in the general framework of geographical thought and the guidelines issued from the "summit" of the profession in the latter half of the 1960s compared with the 1950s are nothing short of remarkable. In his programmatic statement of 1966 "The Past and the Future of Geography"[11] Gerasimov proposed supplanting the traditional systematic specialities of geography with a set of integrated, synthetic disciplines focused on specific problems and regions, with a greatly strengthened economic component. He does not explicitly use Saushkin's image of the "economic conductor of the geographical orchestra" but this is clearly implied in the way in which he presents his "problem areas." He proposed such an organization for the International Geographical Congress at New Delhi in 1968, and it was indeed interesting to hear such a focus on integrated problems presented by Gerasimov at that congress as typifying "the Soviet approach to geography." This new formula for a "constructive geography" has been reiterated and elaborated by Gerasimov in several later programmatic statements,[12] increasingly emphasizing the necessity for an integrated approach to physical and economic phenomena within specific regions and the primacy of economic evaluation of natural resources and locational patterns. Similarly, in two articles originally published in 1965 Kalesnik makes, among other things, a reasoned and careful plea for synthesis of physical and economic geography and for comprehensive studies of regional geography.[13]

It is difficult not to make the connection between such notable revisions of emphasis on the part of these influential leaders of the geographical profession and the various pronouncements by Ilyichev, the Presidium of the Academy, and in *Literaturnaya Gazeta,* as detailed above, or between the latter and the intensive discussion provoked by Anuchin's book. Links in an intellectual chain are always hard to establish with any certainty — and both these leaders still find it necessary to disclaim any connection between their latter-day pronouncements and those of Anuchin and his supporters — but the coincidence is there nonetheless. The prevailing doctrine and recommended guidelines of the later 1960s were so basically different from those of the later 1950s that an observer could be forgiven for thinking that official Soviet geography had experienced a thorough-going revolution — in theory and practice.

Of course worldwide trends in consciousness and opinion, notably the upsurge of the environment or ecology movements, have provided fertile soil for a new emphasis in geography. Anuchin's book happened to foreshadow this development rather clearly, and therefore received a boost from the course of events, so that its role as a catalyst should not be exaggerated any more than it should be discounted. Whatever weight may be accorded to the various operative processes, a perusal of the recent theoretical and programmatic statements from all parts of the Soviet geographical spectrum leaves one in no doubt that the major conflict has been resolved. The situation has reverted to a more normal one for the Soviet Union, with argument largely directed to means rather than ends.

SOME COMMENTS ON THE BOOK ITSELF

Even though the translation of Anuchin's book renders largely superfluous a comprehensive interpretive review of it[14] some comments on some features that might sound ambiguous to Western ears may be in order.

The first half of the book, devoted to a critical survey of the history of world geographical thought and practice, was distinguished not only by being the first such survey of any depth and originality in the Soviet period but also by being markedly more objective in its assessment of non-Russian geographers than had earlier been customary. Such formerly excoriated men as Ritter and Hettner, for example, were by and large treated seriously and at length, which required a good deal of courage, though tenets attributed to others, like Ratzel, Mackinder, and Taylor, were notably wide of the mark (though perhaps hardly more so than some Western stereotyping of them). Apart from one or two Marxist or pre-Soviet Russian geographers, Anuchin lavishes the most unrestrained praise upon Elisée Reclus, which is not surprising considering his combination of revolutionary activities and comprehensive regional studies, together with his being the first to use the concept *geographical environment* to accord with Anuchin's connotation of it.

Anuchin's apparently equivocal use of the term *geographical determinism,* particularly with reference to non-Soviet geographers, needs some explanation. Like some other terms, it should be viewed in relation to the nomenclature of Soviet discussions and notably Anuchin's own tenets and philosophical mission. We should start by considering its opposite "indeterminism," which may be quite unfamiliar to many readers and against which Anuchin has long been concerned to direct his critical fire. He feels that "determinism reflects the view of materialist philosophy" referring to a division basic to Marxist theory. He brands as indeterminists both physical and economic geographers who ignore or deny what he regards as the strong mutual connections between environment and human society. Thus it has been a key word in the fight against the doctrine of a geography divided sharply into physical and economic segments, and also on the practical plane against those planners who hold stereotyped views, rather than those based on detailed study about the character and potential of natural environments. Second, he realized that the specter of geographical determinism had been used as a powerful argument with which to denigrate the whole idea of a unified geography and its protagonists, both foreigners and, by association, Russians, and was therefore a crucial target to try to shatter. Thus we find him counterposing relatively "progressive" geographical determinists or "instinctive materialists," including most of the big names of pre-1920 world geography, with those "God and Tsar" indeterminists, whose feet were not planted firmly on the earth.

Similarly Anuchin's occasional vehemence against the idea of giving location a central place in geographical theory, for example, "The definition of economic geography as a science of location of economic activities seems deeply erroneous to us," needs to be related to certain historical developments, also "leading to indeterminism." In particular this derives from what was called the "branch-statistical" school of economic geography that, earlier this century, apparently played down the role of the natural environment, focused on location, and came close at various times to being transferred bodily out of geography into economics. This so-called leftist tendency kept reappearing and apparently militating against the development of an integrated geography, right up to the late 1950s when economic geography was submerged in a tract on "Economic Science" in the all-important *Great Soviet Encyclopaedia,* and geography was reduced to physical geography alone. The importance of "reading between the lines" in Anuchin's book, and accepting the possibility of special meanings ascribed to these and other concepts, underlines the desirability of knowing something of the context of the history of Russian geography as well as of Marxist theory.

SIGNIFICANCE FOR WORLD GEOGRAPHY

Anuchin's main theses are developed insistently throughout the book, and need not be amplified here. However a few comments seem to be called for by their apparent relevance to many of the urgent issues that are liable to remain with us at least throughout the seventies, both within the academic discipline of geography and in the world at large.

The central, all-pervading object of study of geography is, according to Anuchin, "the geographical environment," which might be equated with "humanized nature" or even, in essence, with the "cultural landscape" (though certainly without, in the latter case, any connotation of restriction to what is "visible"). It is intentionally broad and loosely defined. It has to be seen as a strong reaction and corrective to a Soviet geography that had become, by the mid-1950s, heavily concentrated on the study of a supposed exclusively natural environment with a minor economic geography segment divorced from it and often under the wing of economics. In some ways the situation resembled that of the early 1920s in America following the period of the dominance of Davis's rather self-contained geomorphology, coupled with an environmentalist assertion of natural controls over human history and society. The reaction can be seen, for example, in

Sauer's "Morphology of Landscape" and Barrows's "Geography as Human Ecology,"[15] restoring primacy to man in the geographical equation, emphasizing the importance of his influence on the natural environment and leading to a steady decline in the study of physical geography for its own sake (within geography) in America. An important difference is that Soviet physical geography, focused as it is on bioclimatic principles, is basically more amenable to functional integration within geography as a whole than was the American physical geography of Davis's time. Recently the revival of interest in biogeography and such climatic themes as heat-water balance, together with the revival of interest in George Perkins Marsh and the "ecology revolution," have made American geography potentially sensitive to the all-embracing idea of a geographical environment as set out by Anuchin.

Closely entwined with this idea is Anuchin's insistence on a comprehensive integrated approach to geographical problems, and the belief that it is in regional geography that "the geographical method probably finds its fullest expression." Here again the background was one of a long period of neglect of regional studies in the Soviet Union and a diminishing supply of competent many-sided regional specialists, even for the major Soviet regions. On the practical plane (and this is always at the back of his mind) he claims that the neglect of this all-round geographical specificity has done great harm to the economy, and obviously this argument has proved of crucial importance in the eventual winning over of various Soviet authorities to Anuchin's theoretical point of view. Chapter four, for example, is a perceptive statement of the practical dangers of overspecialization and the loss of an integrated overall geographical approach — the kind of argument that has been increasingly in need of reassertion in the West also over the last decade or so. He is also convinced of the necessity for historical depth in geographical studies and a broadly cultural as well as a practical role for geography — features that have not been conspicuously promoted in Russian geography since the 1920s. Alongside this he has recently voiced his concern that the spread of the use of mathematical models in Soviet geography, much of which he endorses, should not stifle "the art of geographical description."[16]

The spirit and concerns of present-day American geography seem diametrically opposed to those which informed it in the first two decades of the century, when the formal profession was in its infancy and Davisian geomorphological theory and environmental determinism captured the stage. In contrast the character of the Soviet geography that has been ushered in during the decade following the appearance of Anuchin's book has much in common with the spirit of the Russian geographical traditions before 1930. So much so that it now seems justifiable to speak of a Russian school of geography (as distinct from, though not necessarily in opposition to, a Soviet school) that shows a meaningful continuity today. The current activities and themes of Soviet geography are very much more similar and relevant to the theory and practice of geography in America and the West generally than they were a decade or two ago. Moreover the controversy has probed and stirred up some general currents of thought in the intellectual life of that country, comparable in depth and breadth with those of the long drawn-out genetics controversy there, with which it overlapped. These are some of the reasons that lie behind the belief that the translation here offered to the English-speaking world should prove significant and thought-provoking to many people concerned not only with the health of an academic discipline but also with matters of life and death in the future of our planet and its inhabitants.

SEQUEL

Shortly after this English edition had been initiated, a revised version of Anuchin's book, renamed *Theoretical Foundations of Geography* (*Teoreticheskie Osnovy Geografii*, Izdatelstvo "Mysl," Moscow, 1972) was published. In its spirit, purpose, and dominant themes it does not differ fundamentally from its original forerunner, although changes certainly have been made in emphasis and in tone. The chief new feature is an extension of the space devoted specifically to practical environmental problems and the significance of the geographical approach in preventing and curing them. Although many of the examples mentioned naturally related to Soviet projects, the discussion is unusually wide ranging and international for a Soviet work. Topics touched on include the encroachment of cities onto prime agricultural land, world population growth, conservation and consumption of energy and water resources, the "green revolution," and so forth. Anuchin hammers home his long-standing injunction to planners to take carefully into account local environmental conditions and not close off the most sensible long-term options for the use of particular regions. He even advocates assigning a price to land and natural resources, something that has long been discouraged by Marx's labor theory of value. All this has considerable significance in the Soviet context and should influence Soviet planning circles, especially as Anuchin himself has now resigned his professorship at Moscow University to take up a position with the "Council for the Study of Productive Forces," a think tank for long-term national planning.

The bulk of the new book is still focused on the history of ideas and methodological and philosophical analysis. The major themes are substantially the same, with particular emphasis on the seamless geographical environment, "saturated with the results of labor," and on the key role in geography to be accorded to comprehensive regional studies. However, the general tone is somewhat less strident, and the judgements on the domestic opponents as well as foreign geographers more magnanimous, which is hardly surprising, since in the decade between the appearance of the two books, the ideas of the rebel had become accepted, albeit tacitly and often reluctantly.

Interesting though Anuchin's new volume is, it in no way diminishes the importance of publishing an English edition of the original book. Apart from the intrinsic interest of the latter's message and tone, its significance is enhanced by its role, now recognized, as a key document in the history of Soviet geographical thought. Presumably an extended world readership for it should increase the interest in the sequel and perhaps build up a demand for its translation as well, at least the latter part of it. But in the history of ideas, it is always more crucial to have access to the original document — with all its imperfections — which has sparked a grand disputation and provided the impetus for major changes in theory and practice, than to a revision issued when these changes have been absorbed by the intellectual establishment, and thereby taken out of the arena of controversy.

1. See, for example, the report by A. A. Grigoriev, then director of the Institute of Geography of the Academy of Sciences, on "The State of Soviet Geography" in the London *Geographical Journal* for 1955.

2. See David Hooson, "The Development of Geography in Pre-Soviet Russia," *Annals of the Association of American Geographers* 58 (June 1968): 250–72.

3. See, for example, A. A. Grigorev's article, "The Geography of the Capitalist Countries in the Era of Imperialism," in volume 10 of the *Great Soviet Encyclopedia* (1954), which happened to be the only Soviet publication listed in Hartshorne's *Perspective on the Nature of Geography* (1959).

. *Soviet Geography: Review and Translation,* October 1961.

5. *Geografia v shkole* [Geography in School], no. 4 (1961), p. 90.

6. *Soviet Geography: Review and Translation,* September 1962, p. 11.

7. Ibid., December 1961, p. 17.

8. Ibid., October 1963, p. 30.

9. Ibid., April 1964.

10. All are published in *Soviet Geography: Review and Translation,* September 1965.

11. Ibid., September 1966.

12. See, for example, ibid., November 1968 and April 1971.

13. In ibid., September 1965 and September 1966.

14. See my review in the *Annals of the Association of American Geographers,* December 1962.

15. Carl O. Sauer, "Morphology of Landscape," University of California Publications in Geography, vol. 2, no. 2 (Berkeley: University of California Press, 1925), p. 55 ff; Harlan Barrows, "Geography as Human Ecology," *Annals of the Association of American Geographers,* vol. 13 (1923), pp. 1–14.

16. *Soviet Geography: Review and Translation,* February 1970.

Since this introduction was first written, two somewhat related books have appeared: A. G. Isachenko, *Razvitie geograficheskikh idei* (Development of geographical ideas) (Moscow, 1971); and I. Gerasimov, *A Short History of Geographical Science in the Soviet Union* (Moscow, 1976). Although both of these are interesting and contain much factual and biographical material, both, like Anuchin's second book, noted above, appeared after the transformation of Soviet geography and do not address themselves substantially to Anuchin's original book and the controversy surrounding it.

PART TWO

THEORETICAL PROBLEMS OF GEOGRAPHY

SOVIET PUBLISHER'S NOTE

V. A. Anuchin's book, *Theoretical Problems in Geography,* is of a controversial character. The author is an advocate of the concept that modern geography is a unified, integrated science with a single object of study. However, the author does not rule out, but rather affirms, the internal division of the science into physical and economic geography, considering this division to be relative and emphasizing the common elements that unite the two branches into a single complex. To substantiate his viewpoint, the author examines in detail a number of basic theoretical aspects of geography (the concept of geographic environment, its composition and structure, its relationship with society, the content and objectives of regional geography and its place in geography, and so forth). In this connection the book includes a brief historical review of the philosophical doctrines that have served at various times as the basis for the construction of the concepts of determinism and indeterminism in geography.

The author gives considerable attention to problems that are of specific importance for economic geography. In particular, he examines in detail the existing definitions of economic geography as a science, its subject matter and methodology, and its interrelations with the various economic sciences.

The author attributes the present sharp differentiation between the two branches of geography — physical and economic geography — to its unbalanced development (differentiation alone, without adequate development of integrated inquiry) and the same sort of one-sided trend (very narrow specialization) in the training of geographers.

The author believes that this differentiation impedes the development of the socialist national economy, which at present, during the building of a material and technical base for communism, especially needs to supplement research in specialized disciplines with integrated inquiry into the geographic environment and productive forces revealing their close ties and interdependence. Proceeding on this premise, the author attaches special importance to regional geography. However, his arguments on this point are inadequate.

The author calls the concept of an integrated geography monistic, whereas he terms dualistic the concept that physical and economic geography are autonomous sciences. One can hardly consider well-advised or fortunate the transposition of these terms from philosophy, where they have a very definite meaning, to geography, where the author gives them a completely different interpretation.

The publisher is printing V. A. Anuchin's work in the belief that its basic theses and proposals are of definite theoretical and practical interest.

The book is aimed both at geographers of all backgrounds and at specialists in philosophy, history, economics, and natural history.

AUTHOR'S NOTE

The content of this book is based on the author's desire to substantiate a monistic, materialistic view of geography. To do so, it was necessary to become familiar with the history of the basic ideas that have determined the developmental trends of geography from its origins to the present.

The first three chapters are the result of that familiarization with the history of ideas in the field of geography. They are by no means intended to be a detailed or systematic exposition of the theoretical and methodological theories that have been popular among geographers. The extremely limited factual material and examples from the history of geographical ideas are cited in order to show the basic *opposing* trends in the development of geographic theory and to show the "front lines" in the struggle of opinions that have been waged in geography. The remaining chapters set forth the author's system of views on geography. They also show a struggle of opinions, but mostly in its present-day manifestations.

The book engages in minimal debate with individual representatives of the various schools of geographic theory. It does not contain a detailed critical survey of these schools. The criticism that is included is directed not against individual opinions (or their adherents) but exclusively against concepts that deny the unity of geography. The criticism of individual works is also based on a desire to emphasize certain positive theses.

The absence of a monistic theory of geography; the popularity of theories that confine integrated research only to groups of either purely natural or purely social components of the geographical environment; the denial of the scientific content and importance of regional geography and the assertion that it is something external to geography; the relegation of earth science completely to natural science — all these notions, based on a denial of the unity of geography, lead in their totality to a denial of the possibility of perceiving the geographical environment as a definite entity.

Meanwhile, new demands on the utilization of the geographic environment are arising in the socialist countries, with their new and higher forms of public economy. The new social relations allow the geographical environment to be put to practical use in an integrated manner, which in turn opens up great possibilities for the further growth of productive forces. For this reason social development in the socialist countries is beginning more and more to experience a need for synthetic geographical works that depict the entire complex of natural and economic conditions of a territory. And this, in turn, imparts practical significance to research of a broad geographical character, including general geographical research.

The author, after spending a number of years involved in integrated geographical research and trying, on the basis of it, to produce concrete regional-geographical (in other words, general geographical) works, has become convinced from personal experience that the development of geography and its application in economic practice are greatly hampered by the absence of a monistic theory of geography. The author's experience has demonstrated to him that as long as incorrect views of the unity of geography are widespread, general geographical work cannot be developed successfully.

It is also perfectly obvious that geographers must now choose one of two trends for the further development of their science:

1. One may argue that geography does not exist as a unified science, and that as a result of its differentiation there now exist only separate types of geography (two or more), developing as related but distinct sciences. The conclusion to be drawn from this thesis is that there is no possibility of any work of a general geographical character.

2. Or we must recognize that differentiation is merely one

aspect of the development of geography and does not destroy its unity. Then one must recognize that some branches of geography are of a social character, whereas others belong to natural science.

The fact that the absence of a well-elaborated theoretical concept of geography as a science is one of the serious obstacles in the path of its development is becoming a universally recognized truth. It is not by chance that in recent years interest in theoretical research has increased noticeably among Soviet geographers, and more works of a theoretical character have begun to appear. In our view, the works of N. N. Baranskiy, N. N. Kolosovsky, I. S. Shchukin, Yu. G. Saushkin, and V. N. Sementovsky are of especially great value for the creation of a monistic concept of geography.

Without claiming at all to offer an exhaustive solution of the basic theoretical problems of modern geography, the author hopes his book will promote, at least to a small degree, the creation of a monistic concept of geography.

In conclusion, it must be noted that the appearance of this book would have been impossible if its author had not been aided continually by a group of geographers at Moscow State University, where the book's sections were discussed a number of times. In expressing his gratitude to this group, the author would also like to express special and profound gratitude to his teacher, N. N. Baranskiy, who examined the manuscript of this work and provided a great many pieces of valuable advice and criticism.

The author is also very grateful to the collective of the Moscow Branch of the USSR Geographical Society, where the manuscript was the subject of a thorough discussion at a joint meeting of the departments of economic and physical geography and was recommended for publication.

CHAPTER ONE

The Origins of Geography and
the Two Trends in Its Development
in Ancient Society. Synthesis Under
Conditions of Inadequate Concrete Research
and Insufficient Ties to Applied Work.

Man started to acquire geographical knowledge in early antiquity, evidently even before the emergence of the slave-
owning system, since even the most primitive economy was
impossible to manage without such knowledge. Although geographical theories varied among different tribes and peoples
during these times, one can still see in them several important
common features. The tribes and peoples of early antiquity
thought of their place of habitation, their country (or more
accurately, habitat) as being the center of the world. Their
concrete geographical knowledge was distinguished by its territorial limitations. Although they had adequate knowledge of
the territory of their settlements and the environment in which
they struggled for existence, primitive peoples knew very little about the territories beyond their own.

Observations of nature by primitive peoples reduced to the
establishment of individual facts, without ascertaining the
general character of territories. Therefore in our modern terms
these observations could scarcely be called geographical. The
processes occurring in nature were perceived as the acts of
gods and demons. Still, even before the emergence of the
slave-owning system, ancient peoples had a store of knowledge about rocks, plants, animals, winds, sea currents, and so
forth. The needs of economic life, trade among tribes, conquests and collections of tribute all resulted, as did many

other things, in the necessity of accumulating concrete geographical knowledge.

It is well known that roving hunters were already drawing
crude but rather accurate map-diagrams. The ancient Vikings
and Polynesians, who made close studies of sea currents,
trade winds, and coastlines, knew how to make maps and carry
out long voyages, using the stars to guide them. We also know
that the drawing of maps on hides was widespread among the
Indians of Labrador, and that coastal maps were made by
Eskimos. Thus geography, like every other science, arose as a
result of practical needs and arose from specific details.

In a study of the history of geography, one must not bypass
the period of its development in ancient society, since the
principles that formed the basis of modern geography had
already been suggested at that time.

The development of geography took a substantial leap forward during the slave-owning period because it was under the
slave system that a sharp delineation between physical and
mental labor occurred, and a vocational and territorial division of labor emerged and took shape.

One can see considerable development of geography in the
most ancient slave-owning states. In Egypt, for example,
maps were used at least thirteen hundred years before Christ.
The ancient inhabitants of slave-owning Mexico also knew
how to draw maps long before the first Europeans appeared
on the American continent. In ancient China, geography,
along with history, may be called one of the earliest branches
of knowledge. It developed there on the strength of applied
needs, involving primarily the development of irrigation. The
earliest farming in the valleys of several Chinese rivers, where
feudal states later formed, would have been impossible if the
population of these valleys had not had a certain amount of
geographical knowledge. Finally — and this is especially important for the purposes of our work — the slave-owning
period produced the first cosmogonic doctrines, which, although they developed within the theoretical system of
natural philosophy, also contained the seeds of geography as a
specific field of human knowledge. This period marked the

beginning of the theoretical comprehension of geographical phenomena.

In a number of countries of the slave-owning world (Egypt, Babylonia, India, China, etc.) more and more frequent efforts were being made to examine scientifically the surroundings of human society. Efforts were made to create scientific concepts of the earth and its surface, and many philosophers proceeded from hypotheses involving the existence of a material cause, but one interpreted very naïvely (if one approaches these hypotheses from the standpoint of modern science).

The first theoretical concepts of a geographical character developed *within* the cosmogonic, often basically materialistic, hypotheses that were contained in the doctrines of the philosophers of slave-owning society. The struggle between materialism and idealism, which began in earliest antiquity, was reflected in the level of knowledge about the earth and the level of geographical concepts.

Some of the more prominent countries of the ancient world were those of the Mediterranean, where the slave-owning system of production was most highly developed and where, because of this, the development of science and culture, although it started a little later than in some Eastern countries, attained its highest level. The rise of these countries to the ranks of the most developed was also aided by the geographical factor, particularly the territorial differentiation of natural conditions, which was of great importance for social development in the early stages of human history.

In the Mediterranean, at the juncture of three continents, fertile river valleys alternate with arid deserts; the shoreline either cuts deeply into the continents, forming numerous gulfs, or runs almost without bends along marshy coastlands. In some places the sea is dotted with numerous islands, whose archipelagos seem to form bridges between the continents, whereas in others hundreds of kilometers of water separate opposite shores. The mineral-rich mountains either come close to the seashore or run inland, making room for coastal valleys. Desert lowlands alternate here with beautifully forested mountain areas. In short, probably nowhere else in the world could one find an area where such differentiation in geographic environment existed on a territory of comparable size. One should therefore not be surprised that it was in the Mediterranean that one of the oldest and most important cultural centers emerged, and the development of productive forces acquired the greatest diversity.

Here one can very well apply a definition of Marx's that says that "... not the absolute fertility of the soil but its differentiation and the diversity of its natural products constitute the natural basis of the social division of labor; thanks to the changes in the natural conditions in which man must conduct his economy, this diversity contributes to an increase in his own needs, abilities, means and methods of labor."[1]

Ancient Greece, which inherited many achievements from other Mediterranean peoples, had a highly developed culture. Efforts were made in that country to understand man's environment and to establish ties between that environment and social life; the concrete geographical knowledge that existed at the time was systematized to some extent. In the process, the Greeks borrowed a great deal of knowledge, especially geographical, from the Eastern peoples, whose cultural development began earlier. For example, the Greeks were greatly influenced by Phoenician culture. Phoenicia occupied a narrow eastern Mediterranean coastal strip bordered by the Lebanon Mountains. The barren soils and the absence of large territories made impossible a diversified development of agriculture. It was limited chiefly to the cultivation of olive trees and date palms. The population's chief occupation was initially fishing, and later maritime trade, which was occasioned by the growth of countries whose economies were based on irrigated farming, primarily Egypt and Babylonia.

Ancient Phoenicia was a conglomeration of individual slave-owning city-states that gradually took over a considerable portion of the Mediterranean coast through their city-colonies. Every such city-colony was sure to have a harbor that was well fortified by the standards of the time. Particularly important among them was Carthage.

The Phoenicians transformed the open sea into a means of contact between peoples. While engaging in trade (including slave trade), they frequently combined trading operations with pirate raids intended mainly to capture slaves. The Phoenicians sailed in their galleys[2] to the western shores of Spain and France and were the first Mediterranean navigators to establish ties with Great Britain, where they even organized tin production. The Phoenicians also sailed to the North Sea and possibly reached the Baltic.[3] They penetrated far to the south along the western shores of Africa and probably even crossed the equator. Information about India also came from Phoenician sailors.

A considerable part of Greek culture was based on the achievements of the Phoenicians; it was from the Phoenicians that the Greeks took their alphabet, and a large number of words related to maritime navigation, trade, handicrafts, farming, and so forth.[4] The organization of the Greek slave-owning society itself, which arose at a certain stage of the development of productive forces, was in many respects reminiscent of the organization the Phoenicians had. Greece, for example, like Phoenicia (and, incidentally, like many other slave-owning states), was for a long time a conglomeration of individual city-states.[5] The Greeks, like the Phoenicians, sent surplus free people to sea for trade and piracy. In the process new maritime city-colonies were founded far from their country.[6]

This kind of organization of the Phoenician, and later of the Greek, slave-owning society unquestionably contributed to the accumulation of geographical knowledge. The Phoenicians, and later the Greeks, had fairly accurate data and correct conceptions of the Mediterranean Sea and its coast. The inhabitants of coastal city-colonies had some knowledge of the countries inland and of the peoples populating them.[7]

The final separation of mental from physical labor was a consequence in Greece of the political revolutions in the sixth century B.C., which established the slave-owning system. It should be pointed out here that slavery, on which the economy and culture of ancient Greece were based, was at that time a new and higher level of social development that broadened the material foundation of society. Slavery promoted an increase in the production of goods, an expansion of farming, cattle-raising, and handicrafts and thereby promoted the development of trade. The slave-owning states created caravan and maritime trade. Commodity and monetary relations developed. Cities grew into centers of economic, political, and cultural life. The cruelly exploited slaves, by producing food and commodities, ensured the accumulation and concentration of riches in those cities, which strengthened the material foundation of the culture developing there and enabled philosophy, the natural sciences, and the arts to emerge and develop.

The slave-owning class, which established its political dominance, had progressive ideologists who were interested in the development of productive forces and turned to nature, trying to understand its laws. Many ancient Greek materialist thinkers, engaging in a struggle against the mythological view of the world inherited from the period of tribal society, created cosmogonic hypotheses based on natural science that contained the first concepts of geography as a science of the earth's surface.

The teaching of the philosophers of the so-called Milesian school was of great importance in the development of the materialist view of the world. It may be considered that Thales, Anaximander, and Anaximenes, philosophers of the school, created the first theories of geography that were of great importance for its development as a unified science.

Thales (624–547 B.C.) was possibly the first scholar who, based on the rudiments of scientific knowledge, posed the question of what the world consists of and what its real foundation is. He and his followers created a theory of nature as objectively existing, unitary matter that is in perpetual motion and does not need any supernatural basis to explain it.

Thales saw water as the basis of all natural phenomena, visualizing it as a moist foundation in which everything dissolves and from which everything is formed. He imagined the earth as being an island floating in an enormous world ocean (in the manner of the islands in the Aegean Sea). Thales' doctrine on the unity of nature and his conception of the universe were undoubtedly among the first attempts at a materialist understanding of the world. He also posed and tried to answer questions directly related to geography as a science of the earth. For example, he saw the cause of earthquakes in the fact that the moisture in the depths of the earth was capable, when in motion, of causing tremors of the earth's surface. Despite all the naïveté of this explanation from the standpoint of modern science, one cannot fail to acknowledge that it was a materialistic and scientific one for its time.

Thales' theories about the earth, about spatial relations between continents, about the character of the Mediterranean shoreline, and so forth were probably the most accurate of their time. This was no accident, since Miletus at that time was participating very actively in the colonization of the Mediterranean and was carrying on extensive trade, especially with the countries of the East. Thales himself, according to a number of sources, was a merchant and traveler.

Thales' pupil Anaximander (610–546 B.C.) was the first Greek scholar to draw a map of the world known to the Greeks. He is therefore sometimes called, not without reason, the first geographer. Anaximander conceived the bold cosmogonic hypothesis that the universe was created without intervention by gods. He imagined the earth as being a cylinder freely suspended in the middle of the universe. However, in composing his map he adhered to the deeply rooted concept of the earth as a disc (an obvious concession to the most widespread views). The center of the disc contained an inner sea that was connected to the outer ocean.

Anaximenes (582–525 B.C.) advanced the theory that the earth was a wide, flat plate resembling a flying wooden board and floating in air, the primary element, with the sun revolving around the ends of the plate. In contrast to Thales, he saw the material basis not in water but in air.

Thus, the materialist philosophers of ancient Greece attempted to reduce all the heterogeneity of the world to some single substance. In seeking to find a central basis, they could not get rid of their concrete, material conception of matter and were unable to comprehend it as an abstract category.

One can trace the development of the materialistic trend

starting with the representatives of the Milesian school in the philosophy of ancient Greece. The major thinkers who developed the Milesian school's legacy were Heraclitus of Ephesus (530–470 B.C.), the founder of ancient dialectics, and Anaxagoras (500–428 B.C.). Ancient Greek science culminated in the work of Leucippus (500–440 B.C.), and later Democritus of Abdera (460–370 B.C.), who created a hypothesis on the origin of worlds that was brilliant for his time and also advanced a theory of atoms, in which he ascribed all of the world's bodies and phenomena to combinations of atoms.

Democritus's geographical views may be considered the acme of the knowledge gained in ancient Greece about the earth as a planet. Although we know only the titles of Democritus's geographical works and we can read only a few fragments from them, this is enough to become convinced of the great breadth of his scientific interests in the field of geography. Democritus may be considered one of the first scientists of the ancient world who combined a study of the earth as a whole with a study of individual countries. In doing so, Democritus believed that the face of the earth was continually changing. The earth was originally moist and siltlike, then it began to dry and thicken. Here Democritus repeats in a somewhat altered form a hypothesis advanced before him by Xenophanes of Colophon (565–473 B.C.), who contended that the earth had long been under water, which is why shells were found on land, and often not only far from the sea, but even high in the mountains. Xenophanes said imprints of fish and seals had been discovered in quarries in Syracuse and imprints of many maritime animals had been found on the island of Malta; hence he concluded that the earth had been covered with water and sometime would again be under water.

The geographical works of Democritus included a map of the earth and a map for maritime navigation. He did not confine himself to speculative conclusions; his works reflect his own numerous travels and summarize geographical information obtained from many travelers. Democritus has references to India, which he might have visited. This is probably the earliest mention of India in European sources.

Democritus took note of many highly important problems of geography, some of which remain unsolved to the present day.[8] At the same time, Democritus's works mark the end of a highly important period in the development of ancient geography, when it was very closely linked to the philosophic cosmogonies and cosmologies of naïve materialism.

Thus, it was in ancient Greece that the foundations of a materialistic view of the world were laid and the first steps made toward creating a theoretical conception of geography. It should be noted, however, that analogous materialistic philosophic constructs, on which the seeds of natural science were based, appeared even earlier in the countries of the East, especially in ancient India and in ancient China. Astronomical observations were being taken and geographical maps and descriptions were in use in these Eastern countries several centuries before Christ.[9]

It should be noted that the ancient Chinese philosophers primarily developed sociological and ethical views. Even from Confucius we do not have any explanations of the structure of the universe or hypotheses on the origins of all things. Chinese philosophy in general had many features specific to itself; but this specificity does not mean there are no common aspects in the basic patterns by which philosophy developed as a whole. Chinese philosophy was also rich in traditions of a liaison between materialism and dialectics, and in its development the materialist and idealist schools for many centuries waged a struggle against each other that was often accompanied by mutual influence.

The fact that we call the ancient Greek philosophers the first representatives of a materialist view of the world is based on their great role in the subsequent development of European science. It is not the task of this work to examine the history of the origin and initial development of geographical knowledge or to establish the primacy of certain peoples in creating concepts about the structure of the universe, about the form

of the earth and the character of its surface. The only important point for us is that the first attempts to create theoretical
concepts about the earth and the seeds of the sciences concerned with the earth existed in the earliest periods of the
slave-owning system.

In addition to everything that has been said, one cannot
omit the fact that the ideologists of slave ownership and the
philosophers of antiquity showed contempt for any kind of
detailed inquiry, especially that which had immediate practical significance. They very often limited themselves almost
exclusively to speculative constructs. The division of mental
and physical labor that transpired during the slave-ownership
period — a very progressive occurrence for that time — was
gradually transformed, with the development of class contradictions within the slave-owning society, into its opposite,
producing, in particular, a gap between science and practice.

It would be wrong, of course, to speak of ancient science
and practice as being completely separate. The history of
mankind has never seen such a separation. But the specific
character of science in the slave-owning society unquestionably lay in the great predominance of general, speculative
theories over concrete, empirical knowledge. To deal with
science of an applied character and to deal with technology, in
the view of most philosophers of the ancient world, was considered something unworthy of a real scholar. "Plutarch, in
mentioning the inventions made by Archimedes during the
Roman's siege of Syracuse, found it necessary to *pardon* the
inventor: It was improper, of course, for a philosopher to
engage in things of that sort, he reasoned, but Archimedes
was excused because of the extreme situation his fatherland
was in."[10]

One example of the relative separation between science and
practice was the activity of the school of philosophy known as
the Pythagorean school (second half of the sixth century and
the fifth century B.C.), which was dominated by idealistic
theories. The Pythagoreans regarded as the basis of phenomena not a material foundation but numbers, which supposedly formed the cosmic order. Therefore, understanding

the world, according to their concepts, meant understanding the numbers that governed the world.

Despite their development of idealistic theories, the Pythagoreans made a notable contribution to science, since their doctrine of numbers was one of the first attempts to single out quantitative aspects in the phenomena of nature. As V. I. Lenin observed, the Pythagoreans interwove fantastic religious-mythological fictions with the seeds of scientific thought.[11] The Pythagoreans unquestionably made a notable contribution to the mathematics of their time (especially geometry). The important thing for us in this work is to stress that they proposed the idea that the earth was spherical in shape and that the earth's surface was divided into zones. Moreover, they spoke not only about the earth as being spherical but about its motion around a central fire (Philolaus), and later Pythagoreans (fourth century B.C.) replaced the central fire ("celestial hearth") with the sun.

However all these guesses were of a purely speculative character. The zonality on the surface of the earth, the Pythagoreans taught, was merely a reflection of the five celestial zones; thus, in their theories about the earth, the Pythagoreans combined correct ideas with idealistic fantasy.

The idea of the earth's spherical shape was accepted by the greatest scholar of antiquity, Aristotle (384–322 B.C.), but he did not confine himself to speculative theories; he cited evidence to support this thesis. In conceiving of the earth as a sphere, Aristotle deduced that the amount of illumination and heat varied along the earth's surface, depending on the angle of the sun's rays. Hence he determined the existence on the earth of climatic zones.

Aristotle wrote an entire book, entitled *Meteorology,* on atmospheric phenomena, and in it touched on many problems of general geography. Aristotle regarded the sun and especially the effect of its moving closer to and farther from the earth as the original cause of all the phenomena occurring in nature. In another work, *Politics*, he tried to attribute to the influence of climate several features of social relations, and the national character of individual peoples in particular. Here he was close to the ideas formulated before him by Hippocrates (460–377 B.C.), who regarded people's characters as being derived from climate, and by Plato (427–347 B.C.), who tied people's spiritual lives to the influence of the sea.

Thus, attempts to attribute social phenomena to the direct influence of natural conditions already existed in the philosophy of ancient Greece.

In noting Aristotle's services to science, one should also recall that he placed the earth in the center of the universe; in his conception, the sun and stars revolved around the earth. In this question Aristotle took a step backward by comparison with the Pythagoreans and, through his authority, established for a long time to come a misconception about the relative positions of the planets.

Aristotle's pupils also included scholars who dealt specifically with geography. One of them, for example, was Dicaearchus (326–296 B.C.), who defined more precisely the doctrine of latitudinal zonality. It was he who introduced the Arctic Circle, the temperate zone, and the equator. In addition, he measured the altitudes of Greece's major mountains by means of a primitive theodolite and wrote a detailed description of most of the Greek islands.

The all-encompassing philosophic system of Aristotle, who devised the ancient world's first classification of sciences, marks the end of the classical, Hellenic period in the development of sciences and philosophy in ancient Greece and ushers in the next, the so-called Hellenistic period, which began at the end of the fourth century B.C. and lasted roughly until the middle of the second century B.C. (when Rome's conquest of the Hellenistic states started). Whereas in the Hellenic period the sciences developed, as a rule, in inextricable unity with philosophy and within it, in the Hellenistic period the degree of development of the sciences, especially the natural sciences, was so high that some of them (particularly mathematics and astronomy) had begun to develop outside of philosophy. Naturalists emerged who were not philosophers, although they, of course, shared the views of certain schools of philosophy. It was in this period that geography began to stand apart from philosophy.

The most prominent representative of geography at that time was Eratosthenes of Cyrene (276–194 B.C.), who lived in Alexandria, where for a long time he headed a famous library. A broadly educated scholar and poet, Eratosthenes had at his command most of the scientific knowledge of his time. His best works are devoted to geography. Eratosthenes was the first to undertake specialized geographical research, and he used mathematical methods. He sought to create a scientific basis for geography, for which he considered it essential first to establish the dimensions of the earth and to determine accurately the position of the ecumene (i.e., populated part) on its surface. He achieved stunning success in these efforts. The method he devised made it possible for the first time to perform a truly scientific measurement of the earth and to obtain a result that was fairly close to reality.

On the whole, Eratosthenes located the latitudinal zones on the earth correctly, making use of Dicaearchus's works for this purpose and supplementing them. Eratosthenes considered the ecumene to be an island, proceeding from the concept that water exceeded land in surface area and assuming that the length of the ecumene amounted to a little more than one-third of the earth's circumference; he therefore expressed ideas to the effect that there may be other inhabited worlds on the earth that were unknown to the peoples of the Mediterranean. He even pointed out that such a new inhabited world (or worlds) was most likely to be in the temperate zone, extended westward across the Atlantic Ocean. One may consider on this basis that Eratosthenes thereby presumed the existence of America. He also voiced the idea that it was possible to reach India by traveling west, since the Atlantic Ocean, he was convinced, had to be connected to the Indian Ocean. Eratosthenes wrote a voluminous work, comprising three volumes, which he called *Geography*. He began the history of geography with early Ionian maps, and not with Homer, whom he rejected as someone who knew very little and invented very much. This work also set forth the method he used to determine the size of the earth and gave a description of the ecumene.

Unfortunately, Eratosthenes' works, which were far ahead

of the science of his time, did not become universally known, and his discoveries were long ignored.

In addition to the differentiation of geography, its parallel development in two different directions is clearly evident even before Aristotle. One group of scholars gave its attention to studying the earth as a whole; this work was closely coupled with cosmogony and mathematics. Another group engaged in studying individual countries, areas, and localities. This second trend was closely tied to history and essentially may be called a regional trend dealing with countries and peoples.

At the time ancient society was flourishing, the first trend was represented by Aristotle, Eratosthenes, Strabo, and Hipparchus. Questions of general physical geography were examined by the philosopher Epicurus (341–270 B.C.), a follower of Democritus, and by scholars from the late years of the Roman republic, including Cicero (106–43 B.C.) and Lucretius Carus (99–55 B.C.).

The second trend had a leading representative in the person of the historian and geographer Hecataeus (end of the sixth century to the beginning of the fifth century B.C.), who was nicknamed "wanderer" because of his numerous travels. Hecataeus corrected and substantially added to Anaximander's map of the world. Furthermore, he wrote the work *Description of the Earth*, from whose title the term *geography* was apparently derived. This work has not survived, but one can still determine from the preserved fragments that it contained a rather detailed description of a substantial part of Europe, Western Asia, and Libya. A large role in the development of ancient regional geography was played by the "father of history," Herodotus (484–425 B.C.), the historian Polybius (201–120 B.C.), and the outstanding geographer of the ancient world in the initial period of the Roman Empire, Strabo (63 B.C.–20 A.D.).

Perhaps the most striking representative of the regional trend in the development of ancient geography was Herodotus, who included a great deal of geographical material in his historical accounts. He had extraordinary geographical and ethnographic knowledge. For example, he gathered an exceptionally large amount of material about the peoples who populated Africa at that time. Traveling widely and possessing keen powers of observation, Herodotus kept highly detailed descriptions of the lands he visited. Moreover, he summarized the isolated historical and geographical accounts of countries compiled by other travelers.

Herodotus's vivid accounts included descriptions of relief, soil and vegetation, data on climate and animals, details on roads, cities and specific characteristics of the culture and everyday life of the population. Moreover, he pointed out differences within countries based on economic specialization of their individual areas, indicating where the population's chief occupation was trade and handicrafts, and where it was farming or cattle-raising.

Herodotus, as we know, was the first geographer to visit the Greek colonies in the Crimea and the lower reaches of the Dnieper; he may have also visited places somewhat further north. He amassed a great deal of information about the territory and peoples of southeastern and eastern Europe. He knew the Caspian Sea was a large lake and that Scythia extended far to the north. His concepts, consequently, were not marred by the delusions that became widespread later regarding a link between the Caspian Sea and a northern ocean. No other ancient geographer described areas as far north, into the heart of northeastern Europe, as did Herodotus.

Without question, Herodotus's works had a powerful influence on the further development of the regional trend in geography. Although he, like Hippocrates, linked social life to natural conditions, he took a substantial step forward in comparison with his predecessors, who attributed virtually all features of social life to the influence of climate alone.[12] Herodotus believed that the character of peoples depended on many factors, both natural and sociohistorical (culture and historically evolved traditions).[13] Herodotus's views were largely repeated by Strabo. Poseidonius (135–60 B.C.) held that not only plants and animals should be approximately identical at the same latitudes but also people and their social life, whereas Strabo criticized such contentions and spoke of the inevitability of one people borrowing habits and customs from others. On the basis of this feature alone, Strabo asserted, the characteristics and differences in people's personalities cannot be attributed only to climate. Here one can already see an allusion to the importance of the human mind in the shaping of social life.

The travels of Pytheas, Ephorus, and others, and especially the campaigns of Alexander the Great, broadened the horizons of geography and furnished a great deal of new information about countries and their peoples. The success in the development of astronomy, mathematics, and natural-science disciplines made it possible to prove the earth's spherical shape experimentally and on that basis to elaborate a theory of climatic zones.

By the time ancient society was flourishing, much data had been gathered about plants and animals, and the population geography of the parts of the world known to the Greeks.[14] The establishment of an interconnection between man and nature led to the point where human society was viewed not as a special spiritual category but as part of the material world of nature. It was on this naturalistic and materialistic basis that progressive philosophers struggled against religious prejudices.

The differentiation between the two trends in the development of geography also became evident in the following period; but it would be wrong to think that these different and, it would seem, so dissimilar cores of geography were not related at all. It would be wrong if only because the science of that time was unified by philosophy. We have already referred to Democritus, who linked a general concept about the structure of the earth to a concrete study of individual territories. Although Democritus in this respect was far ahead of his contemporaries, one cannot say that he was a solitary exception. Epicurus, Lucretius Carus, and especially Strabo, to name a few, sought to make use of a general concept of the earth and to apply it to a concrete study of countries. Representatives of descriptive geography often sought to gather together the phenomena they observed and to create general pictures of the world from these combined details. Thus, there was already an effort in antiquity to establish a common tie between the study of the earth as a whole and of individual territories.

Still, ancient regional descriptions most often consisted of a collection of data about countries and their peoples. The absence of an adequate connection with general geography, the absence of a general theoretical concept of geography as a science, and the territorial limitedness of isolated observations hindered the creation of synthetic generalizations. It is therefore not surprising that the effort to encompass the whole could not be based completely on science and could not over come the religious and mythological concepts that led away from scientific understanding of the earth. Thus, there arose in ancient Greece the idea that the earth was a living organism made animate by a mystical world soul. This idea was voiced for the first time in Plato's *Timaeus* and then was picked up by some other thinkers, among whom Poseidonius should be mentioned. This scholar and philosopher was known as a historian and commentator on the *Timaeus*, but even more as an astronomer whose research contributed a great deal to science. For example, Poseidonius correctly contended that the sun was many times larger than the earth. But at the same time he injected a great deal of mysticism into general geographical concepts.

The last major representative of ancient European regional geography who also had a good knowledge of all the geographical theories of his time was Strabo. This Greek geographer and historian traveled extensively and wrote a multivolume work in which, besides describing all the countries known at that time, he also set forth general questions of geography. Strabo sought to create integrated images of the world without separating man from nature, which was shown in connection with the people living in it, with their economy and all the characteristics of their life. Although he was opposed, as was mentioned above, to attempts to account for social phenomena by the influence of climate alone, Strabo, like many of his predecessors, did not see the qualitative distinctiveness of human society and attributed social phenomena to the direct influence of the natural environment. For example, he attributed Rome's political-economic successes to the direct influence of Italy's advantageous geographical situation and favorable natural conditions. This misconception continued to be widespread afterwards, since not only ancient philosophy but also the philosophic concepts of subsequent epochs (feudalism and capitalism) could not (and, naturally, did not have the capacity to) give a correct explanation of the unity of the material world without mechanically confusing its individual, fundamentally different forms.

Rome's dominance was based on its numerous military campaigns, especially in the northwest and southeast. The general concepts of geography did not undergo any substantial changes in that period, but the boundaries of the known world expanded considerably. The increased range of concrete knowledge was accompanied by a widening of the gap between general and regional geography.

The Romans adhered to two opposite points of view in their concepts of the earth's surface. Some scholars believed that the earth's outer envelope was covered chiefly by the waters of seas and oceans, which were separated from each other by land masses. Others were of the opinion that the earth's surface was an enormous land mass that contained isolated inner seas. In essence, neither of these theories was original with Roman scholars — they were merely a development of the views formulated by the ancient Greek philosophers.

The broadening of trade ties and numerous military campaigns resulted in the accumulation of new information about lands that previously were almost totally unknown, which substantially increased the Romans' geographical horizons. This increase in geographical knowledge was reflected in the works of the Alexandrian astronomer and geographer Claudius Ptolemaeus [Ptolemy] (90–168 A.D.), who made a map containing almost all of the geographical information about the earth's surface that existed at the time. Ptolemy's works probably represented the most consummate expression of the ancient world's astronomy and cartography. He used a graticule and introduced terms corresponding to the concepts of geographical latitude and longitude. His map went on to serve for many centuries as the chief source for cartographic work.

Ptolemy followed Eratosthenes in singling out geography as an autonomous field of human knowledge. His geography was no longer a set of speculative fragments but an autonomous science closely tied to mathematics and astronomy. Distinguishing two trends in the development of geography — general and regional — Ptolemy subdivided the science of geography into two parts: (a) geography and (b) chorography. He also included mathematical geography and cartography in geography, since he viewed geography primarily as the science of the cartographic representation of the earth.

The difference between geography and chorography, according to Ptolemy, was that geography dealt with the earth as a whole, singling out only the most substantive elements, whereas chorography gave detailed accounts of individual countries with all of their noteworthy features. Geography afforded an opportunity to view the entire earth in a single generalized representation, whereas chorography painted a multitude of isolated, specific pictures for us. Geography paid more attention to quantity, concerning itself above all not with the characteristic of individual countries or localities but with the accuracy of computed distances between them. Chorography had to pay more attention to quality and concern itself with the accuracy of the description of a locality under study. Ptolemy attempted, consequently, to explain and substantiate scientifically the differences that actually existed at the time between general and regional geography by formulating simultaneously the major tasks that, in his opinion, confronted each. This kind of effort was a development of great significance also because Ptolemy viewed geography and chorography as sciences with a common object of study, but sciences that dealt with this object differently: geography, as a whole, and chorography, in detail. This interpreatation of geography was the most accurate one for that time.

As a general conclusion for this very brief survey of the major theoretical concepts in the geography of the ancient world, the following may be said.

Geography, like all the other sciences of the slave-owning society, initially developed within the framework of philosophy. The separation of geography as an autonomous field of human knowledge was at that time in its infancy. The general absorbed the particular. Therefore the development of theoretical concepts of a geographical character was bound up in the most direct way with philosophy. The science of that time was not dismembered into branches. Individual scholars could possess, if not all, then almost all, of the knowledge accumulated by mankind.

The science that developed in the slave-owning society generalized the initial knowledge about the earth, its shape, size, and position in the universe. Certain laws of the development of the earth's surface (e.g., the law of latitudinal zonality) were already formulated in general outline at that time. Finally, ancient geography created the first scientific descriptions of certain parts of the earth's surface, which marked the beginning of regional geography.

The progressive, materialist philosophy of the ancient world established the unity of the material world and the unity between human society and the rest of nature. It waged a struggle against the religious indeterminist views, which separated man from nature. The progressive philosophers of the ancient world established the existence of a general connection between phenomena in the material world, which unquestionably was a major achievement of science having great significance for the development of geography.

The general geographical concepts that arose on the basis of speculation resulted from attempts to create cosmogonic hypotheses. Developing separately, in large part, from the regional study of the earth's surface, the concepts of general geography were unable to grow into a science of the laws of development of the earth's envelope (or sphere) that we now call the landscape envelope.[15]

The second, regional trend in the development of geography was closely bound up with historical descriptions. In writing works that described the geographic environments of individual countries and localities, the representatives of this trend rarely tied these descriptions to the general concepts of the earth that already existed. As a result, ancient regional geography was most often purely descriptive, without any attempts to explain the phenomena and events being described.

Although they correctly understood the overall unity of the material world, the philosophers of ancient times were unable to explain many of its particular phenomena, including those of a geographical character. The concepts of the ancient world's materialist philosophers helped strengthen the idea that man was a part of nature, but they still had very poor knowledge of empirical reality and were able to satisfy society's practical needs only to a very small degree.

Ancient geographers often traced all the features of people's lives directly to the influence of the natural environment, attributing features of social life to the direct effect of the laws of nature. In other words, one can see in the geography of the slave-owning society the emergence of a school that was later given the name *geographical determinism*. It must be stressed here that the determinist school arose and developed as a materialist school and disputed unscientific religious views on a materialistic basis. However, the ideas of geographical determinism were also used sometimes by certain idealists. For example, Plato spoke of the influence of a country's relief and fertility on the way of life and governmental structure of peoples.

However idealistic philosophy as a whole was based on a denial of the material essence both of nature and especially of human society, considering them the expression of a spiritual foundation. Plato taught that the visible world was not the true world and that the essence of things lay in ideas. The material body was always only the temporary shell of an immortal soul. Man and human society were special categories, the earthly embodiment of the spirit. Plato's predecessor and teacher, Socrates (469–399 B.C.), another opponent of materialism, considered scientific inquiry into nature to be something unnecessary and godless. "He denied the natural law-governed character of phenomena in nature. He opposed

Democritus's determinism with teleology (the doctrine of purpose), an affirmation of the primordial expediency of a world controlled by a deity."[16] As a spiritual creation, man was usually separated from nature and was placed in absolute contrast to it.

In the field of geography, the idealistic school manifested itself in creating unscientific theories of the earth as a living, spiritual organism and of nature as an immutable condition created especially for man. It was often proposed that natural phenomena and bodies be studied in strict isolation, especially from man and his life, since he was declared to be a special creation of the gods. True, these views were preached most consistently a little later; but even ancient philosophy, for example, Plato, contained views of this kind. In other words, one can already perceive in ancient philosophy, and consequently in geography, the emergence of a school that denied the causal tie between phenomena and denied determinism; so science was forced to wage a struggle against indeterminism, and consequently geographical indeterminism, as we call that phenomenon in geography.

It was determinism's struggle against indeterminism, as we shall try to show, that was the main front of materialism's struggle against idealism and science's struggle against obscurantism in the field of geography.

CHAPTER TWO

The Period of Empirical Development. Initial Attempts to Create Theoretical Geographical Ideas on the Basis of Bourgeois Philosophy. Determinism in the Geography of the Eighteenth and Nineteenth Centuries.

The demise of the ancient world and the transition to the feudal system in Western Europe were accompanied by a temporary economic and cultural decline. Many cities were destroyed or badly neglected. Handicraft industry declined, and trade started to wane. The centers of life shifted to villages. A great deal of knowledge was completely or partly lost. The culture of the ancient slave-owning society was replaced by feudal narrowness. Even simple literacy became a great rarity in European countries. Theology took the place of ancient philosophy.

Initially, in the first centuries of the existence of feudalism in Europe, the field of geography, like most other sciences, regressed from the level that had been reached in antiquity. The general geographical concepts of Europeans, who were oriented toward the Bible and toward the writings of the fathers of the church, were more naïve and primitive than those of the geographers of the ancient world. The scientific conjectures and the first scientific theories about the earth that ancient science advanced were replaced by views that were patently dominated by fantastic religious fictions.

In the epoch of feudalism the earth was transformed from a sphere back into a rectangular plane, oval, or disc, but with certain changes, attributable to the influence of Christian doctrine. For example, Jerusalem and the Holy Sepulcher began to be placed in the center of medieval maps. Perhaps the most vivid reflection of theoretical concepts in the field of geography may be seen in the world system of Cosmas Indicopleustus [of Alexandria], a Byzantine cosmographer who lived in the sixth century A.D. In accordance with the Bible, he depicted the universe in the form of a parallelepiped with crystal walls, covered by a dome. On the bottom of the structure was inhabited land, surrounded by ocean and having the form of a rectangular plane. The ocean, with four sea-gulfs (the Roman, Red, Persian and Caspian), separated the inhabited land from the Eastern land, where heaven was located and where four rivers — the Nile, the Ganges, the Tigris and the Euphrates — had their source and penetrated the inhabited land.

Yet it would be wrong to think that the epoch of feudalism was totally one of retrogression. This is incorrect, first, because the transition from the slave-ownership mode of production to the feudal method did not entail temporary barbarization in all the countries of the world. This did not happen, for example, in China, and obviously did not happen in many other countries of the East, for it was in the feudal epoch that the peoples of these countries enriched the store of world culture with outstanding achievements.

Second, in the countries of Western Europe as well, where the invasion of barbarians and the replacement of the slave-owning structure with feudalism resulted in the temporary degradation of culture, cities began to grow, interest in science intensified, and scientific thought started developing during the twelfth and thirteenth centuries as a result of the development of commodity production and trade and the improvement of handicrafts.

On the whole, culture took a step forward during the epoch of feudalism. ''During the epoch of feudalism, the ancient slave-ownership culture was gradually supplanted by the new and more developed feudal culture. The peoples of the countries of the East, and then of the West, made use of the achievements of ancient Eastern civilization and of the ancient world and made notable progress in the area of science, technology, art and culture as a whole.''[1]

Geography completely duplicated the path of social development. But the general concepts it inherited from slave-owning society had little in common with practical needs in the conditions of feudal fragmentation and natural economy. Successful development of the economy under the new feudal socioeconomic conditions in the small, separate states required knowledge that enabled improved forms of farming and handicrafts to be created. It was this knowledge, typically, that was adopted first from slave-ownership culture and was developed during the early stages of the existence of the feudal socioeconomic structure. The epoch of feudalism saw an interrupted accumulation of geographical data, especially in the process of intercourse between peoples. The Europeans' conflicts (and ties) with the Moslem peoples of the Near East at the time of the Crusades, for example, helped broaden the Europeans' geographical horizons. We know of the remarkable voyage by the Venetian Marco Polo to China in the thirteenth century. The travels of Moslem (mostly Arab) scholars and merchants also played an enormous role in the broadening of specific geographical knowledge in the Middle Ages.

It should be emphasized that medieval geography developed to the highest degree in the countries of the Moslem East. Beginning in the eighth century, the Arabs established their political rule over a vast territory extending from the Indus to Spain and from the Caucasus to Africa. Their commercial ties covered even more territory. Ancient literature was largely preserved in the East, especially in Alexandria, and this fact helped the Arabs master the principles of Greek science. During this period many astronomical and geographical works by Greek scholars, including Ptolemy's astronomical treatise, *Almagest,* were translated into Arabic. Maps were made and systematic astronomical observations were conducted in the Arabic-speaking countries. The scientists of these countries did not lose their concept of the earth as a sphere; they made degree measurements and compiled astronomical tables. They also wrote quite a number of regional-geographical works dealing with countries and peoples that covered almost the entire populated part of the world known to the Arabs.

The adoption of the achievements of Greek science by the geographers of the eastern Moslem countries occurred also indirectly. The influence of Greek science and culture was also felt through the works of Syrian geographers, which to a considerable extent were based on the data of ancient science (e.g., the works of the Syrian geographer Jacob of Edessa, who wrote in the early eighth [633–708] century). In some cases this Syrian influence was a link between the Greeks and Arabs, and in other cases it was of importance in its own right.

The success of the regional-geographical trend in the Arabs' geography was based on conquests and trade. At the same time, their regional-geographical works, like those of the ancient Greeks, were most often historical as well; historical and geographical material were inextricably tied together. Arab scholars, like the ancients, saw links between nature and society. For example, the fourteenth-century Arab historian Ibn Khaldun (1332–1406) tried to prove that the forces and phenomena of nature influence human society as a constant and powerful factor. The influence of nature, in his opinion, extended not only to people's physical appearance and material life but also to their psychology and their intellectual life. Ibn Khaldun attempted to create a unified theoretical concept of the geographic causality of the historical process. Thus, one can also discern in Arab science the school that we call geographical determinism.

The Arabs' regional-geographical works were most often distinguished by their wealth of content. The accounts of some Arab traveler-geographers (e.g., Masudi and Mukaddasi, who lived as early as the tenth century, and Ibn Batuta, who lived in the fourteenth century) remain valuable to us as highly important sources in the study of medieval geography.

The establishment of ties with Arab culture contributed in large measure to the development of European science and, specifically, promoted an increase in knowledge about the earth. Subsequently, during the fifteenth century, when capitalist attitudes began to develop in the womb of feudalism, and trade started to extend to many countries, the range of geographical knowledge broadened more and more.

In addition, under the influence of practical needs, the development of capitalist attitudes in the small feudal states (particularly in Italy) was accompanied by the appearance of statistics, which was closely associated with geography since it essentially provided data characterizing the economic and political conditions and differences in these states that had evolved in the process of formation of individual countries and regions.[2] Particular attention was paid in this regard to data concerning inhabitants as producers of material goods and as taxpayers. One can see in these statistics the first research of a substantively economic-geographical character.

The end of the fifteenth and the beginning of the sixteenth centuries are called the time of great geographical discoveries. These were decades of rapid development for geography and cartography. In thirty years (1492–1522), from the first voyage of Columbus to the round-the-world voyage by Magellan, man's geographical horizons broadened to include almost the entire earth's surface (not counting the polar regions). The discoveries of countries and lands were accompanied by descriptions of them and relatively accurate determinations of their locations. In this way material was obtained that allowed many scientific theories that had evolved in geography to be substantiated not by speculation but by facts. For example, only after the first round-the-world voyage did the earth's spherical shape become an obvious fact for everyone, and even the church stopped denying it. As a result of the great geographical discoveries, in Engels's words, the boundaries of the old *orbis terrarum* (i.e., circle of lands) were shattered once and for all. The process of the broadening of geographical knowledge is clearly evident in the development of cartography: maps began to encompass larger and larger sections of the earth's surface. At the same time, the scientific concepts of ancient geography were gradually adopted, although sometimes they were rediscovered.

In the late fifteenth century Martin Behaim (1459–1507) created Europe's first globe.

A process of intensive geographical study of nearly the entire earth's surface got under way. Abundant factual material was being accumulated and had to be generalized on a new theoretical basis.

The great geographical discoveries and the establishment of ties with far-off overseas countries gave a powerful impetus to the economy of the Netherlands. That country, which in the sixteenth century occupied a central position in Europe's trade with its colonies, felt a particularly strong practical need for geographical knowledge and, above all, good geographical maps. It was at this time in the Netherlands that there appeared the first geographical atlases in the present-day sense of the term (the atlas of Ortelius, and especially that of Mercator).

The progress of general geography as the science of the earth's surface was a component of the overall progress of natural science and of progressive, essentially materialistic philosophy. The discovery by Copernicus, set forth in his book, *Of the Revolution of Celestial Spheres* (1543), which rigorously formulated the principles of a scientific heliocentric conception of the universe, was reinforced by Kepler, who established the laws of motion of the planets, and by Galileo, who proved experimentally the similarity of the earth to other celestial bodies. Later, on the basis of the success of mechanics — the first of the natural sciences to achieve relative maturity — there appeared the far-reaching cosmogonic conception of the great French scientist and philosopher Descartes (1596-1650), who spent about twenty years of his creative life in the Netherlands. He wrote many philosophic and natural-science works, including *Principles of Philosophy* (1644), which, among other questions, sets forth the mechanics of vortexes and a cosmogonic hypothesis on the origin of the sun, the earth, and other planets of the solar system. Seven years before, in 1637, Descartes had written an equally important philosophic work, *Discourse on Method*, in which he formulated the principles of his rationalist method, which emphasized with enormous force, in opposition to scholasticism, the rights of human reason in the investigation of truth.

The influence of the philosophic and natural-science ideas of Descartes evinced itself above all in Dutch universities. A great many Cartesians began to appear in the ranks of Dutch philosophers and scholars. One of them was Bernhard Varenius (1622–50), author of the book, *General Geography*.[3] This book was the first since the works of the ancient philosophers to offer a detailed theoretical concept of geography, corresponding to the bourgeois stage in the development of philosophy and science. The book by Bernhard Varenius was not only a summary of all that had been accomplished by geography prior to that time, it was also a bold look toward the future. In essence this was the first work since the demise of ancient philosophy to contain a theoretical substantiation for the science of geography, to define the object of study of geography, and to outline the basic methods for studying that object. For example, Varenius used the comparative method, which subsequently was employed on the widest scale in geographical research. Although he regarded the earth's surface as the overall object of study of all geographical sciences (geography as a whole), Varenius at the same time understood the necessity of a differentiated study of it. He saw not only an overall object of study for geography but also specific objects of study for its individual branches. Thus Varenius proved the necessity of developing specific branches of geography — geomorphology, climatology, hydrology, and so forth.

Varenius devised also a program of regional-geographical work, defining the range of questions that should be covered in it. In doing so, he separated the questions related to the natural features of countries under study from those related to the characteristics of their population and economy. He viewed physical and social geography as separate, integrated sciences within the framework of geography.

The range of questions Varenius listed in his regional-geographical works included the following: the geographical situation, size and configuration of the country being described; relief, hydrography, and character of vegetation; characteristics of natural conditions from the agricultural standpoint; minerals and their processing; animals. All this constituted the first part of a regional-geographical work. The second part provided a description of the population with detailed demographic data. Then an account was given of the chief occupations of the inhabitants, their income and the handicrafts that the inhabitants and merchants practice, and the goods that one country sends to other lands. Finally, specific characteristics of culture and everyday life and basic political information.

Varenius's work may serve as a vivid example of the tie between the theoretical concepts of a geographer and a definite philosophic doctrine. The materialistic tendencies of Descartes's philosophy, his teaching of the materiality and infinity of the universe, and the indestructibility of matter were used widely by Varenius. Without the physics of Descartes, the book by Varenius could hardly have appeared at all. Even in his attempts to explain (sometimes unsuccessfully) certain natural phenomena, such as sea tides, Varenius made use of the physics of Descartes, which, as we know, was a highly important element of his philosophy.

The theoretical conception of geography Varenius elaborated in his basic propositions retained its importance for many decades. It is therefore not surprising that his work was translated into many languages, and that the ideas it contained were adopted by progressive geographers of the seventeenth and eighteenth centuries in every country. One can say with complete certainty that the ideas of Varenius acquired currency in Russian geography prior to Lomonosov.

The popularity of Varenius's ideas in Russia is borne out by the works of Tatishchev (1686–1750), whose development of Varenius's theoretical propositions was perfectly obvious. Thus, regarding the division of geography into the general and the particular, Tatishchev wrote: "... such as (1) Universal or general, that is, the whole of the universe, the appearance and size of the whole earth, with its waters, continents, and so forth. (2) Special and particular, wherein only a single area, for instance, France, England, Poland, or Russia, is described with all of its characteristics. (3) Topography, or description of places, wherein a single part of some area is presented, for instance, Saxony, Austria or Bavaria of Germany, or the Great and Little Rus or Siberia of Russia, and even smaller parts, wherein one city with its district is described.''[4]

It is clear from the quotation that, in Tatishchev's view, geography deals, on the one hand, with the earth's surface as a whole, and , on the other, with individual countries and their parts, including cities. This division of geography into two mutually complementary parts that are dependent on each other in their development is one of Varenius's major ideas, and, incidentally, it retains its value today.

However in developing Varenius's ideas, Tatishchev also divides regional geography into two parts — special (regional geography) and topography, which seems to us to be extremely important and correct. In speaking of special geography, which deals with countries, Tatishchev displayed great independence in his thinking, affirming that geographers should write descriptions both of nature and of population and economy. But he also distinguished qualitative differences within geography that arise when it is necessary to turn from integrated descriptions of countries to an intensified study of certain aspects of the population's economic activity. In other words, Tatishchev followed Varenius in the contention that, in addition to the division of geography into parts (general and particular), individual branches should also be distinguished in it: ''Description may be mathematical, physical, and political in quality.'' Consequently, Tatishchev added much of his own in developing the major propositions of Varenius. For example, he gave one of the earliest theoretical justifications of the need to study social elements of the geographic environment; in this respect he surpassed Varenius and his own contemporaries.

Even more was done for the development of geography in Russia by M. V. Lomonosov, who provided theoretical and practical justifications for a number of scientific geographical expeditions. Correctly viewing geography as a science dealing with nature, population, and economy in their unity and with their territorial differences, Lomonosov was one of the first to

advance and substantiate the idea that law-governed interconnections exist between all elements of the geographic environment. It is just since Lomonosov's time that the inter-
connected character of the elements of the geographical environment can be considered proven.

Beginning with Varenius, the entire subsequent development of theoretical concepts in the geographical sciences was
connected with the development of certain highly important
systems of bourgeois philosophy. It is common knowledge
that capitalist attitudes began to develop in the womb of
feudal society during the Renaissance. The basis, pace, and
forms of this development varied from country to country. In
those countries where capitalist attitudes developed earlier,
for example, in England and France, we can speak of the
revolutionary role of the bourgeoisie in the destruction of
feudal relations. Meanwhile, in those countries where
capitalist attitudes developed later, this revolutionary role
was highly relative, and in some countries the bourgeoisie,
although it played a progressive role for a certain period,
never did become a really revolutionary class and quickly
moved into the reactionary camp by forming a close alliance
with feudal elements in society. The reason for this occurrence is obvious. It was that the bourgeoisie of some countries
was late in making its appearance in the social arena and did
so at a time when the numerous and organized proletariat had
already taken shape.

This is why, during the Great French Revolution, the
bourgeoisie of France saw its chief enemies in the first estates
and its allies against the nobility in the peasantry and in the
proletarian elements of the urban population, whereas in a
number of other countries where capitalism developed later,
the bourgeoisie considered its chief enemy to be the proletariat. It was in these countries that the bourgeoisie proved
to be, as V. I. Lenin said so aptly with regard to the
bourgeoisie of tsarist Russia, "the prematurely born child of
history."

These commonly known facts have had to be reiterated
here because the changes in the class positions of the

bourgeoisie in the process of historical development could not help but be reflected in bourgeois philosophy as well, which in turn influenced the character of theoretical concepts in geography.

Here one can discern the following pattern: In those countries where the role of the bourgeoisie in social development was especially revolutionary at a certain stage, materialistic, progressive philosophy developed, sharply attacking the religious feudal view of the world. The ideas of the French enlighteners and materialist philosophers of the eighteenth century revolutionized the social consciousness of their time. They played an enormous role in the ideological preparation of the bourgeois revolution in France.

Meanwhile, in those countries where the role of the bourgeoisie was primarily a struggle for a reformist transformation of the feudal system, materialistic bourgeois philosophy developed far more feebly. The predominant influence in these countries came from idealistic philosophy, which sought to reconcile the development of capitalism with feudal state forms.

Whereas the bourgeois philosophers in eighteenth-century France waged an open struggle against religion and the feudal state while undergoing repressions, the ideologists of the German bourgeoisie were distinguished professors. They were "state-appointed tutors of youth; their works were approved by the heads of leadership; and Hegel's system — the crown of all philosophic development — to a certain degree was even elevated to the rank of the Prussian Kingdom's state philosophy."[5] German philosophy called for the adaptation of bourgeois interests to the conditions of the nobiliary state. In Russia, where the bourgeoisie was even more tardy in its appearance in the social arena, Russian bourgeois philosophy could not even attain the level that had been achieved in Germany. Russia's most progressive thinkers began to search for truth outside the framework of bourgeois ideology.

Such was the distinctive geography of philosophy, so to speak, in bourgeois Europe. France, Germany, and Russia were the three countries where the independence of the development of bourgeois philosophy with regard to the social role of the bourgeoisie was probably manifested in the most pronounced and graphic manner.

The basically materialistic trend in the development of bourgeois philosophy had a powerful influence on geography as well. Some bourgeois philosopher-enlighteners made wide use of the achievements of geography in the struggle to affirm their view of the world against feudal ideology. The so-called geographical school in sociology developed as early as the sixteenth and seventeenth centuries and gathered momentum particularly in the eighteenth century. It gained the most popularity in France. This school of geographical determinism sought to explain all the phenomena of social life through the effect on it of the geographical environment. Although the concepts of geographical determinism must not be confused with the materialistic explanation of history, its orientation against theological dogmas gives grounds to speak of it as being fairly close to valid concepts of the process of sociohistorical development, in so far as it shows a desire to establish objective laws for this development without any supernatural elements. The adherents of geographical determinism assumed there was a causal basis for all phenomena, which was of tremendous importance for a more correct methodological approach to the study of geographical phenomena. By comparison with idealistic concepts, the attempts to explain social phenomena by the influence of the geographical environment, despite all their deficiencies from the standpoint of historical materialism, were of positive value, since they were directed toward discovering the real factors that determined the development of society.

Until Marxism appeared, the representatives of geographical determinism disseminated the most progressive ideas in geography. At the same time, however, the geographical approach in the study of the process of social development and individual social phenomena inevitably led on occasion to unscientific conclusions. Thus, though they saw the organic connection between human society and the rest of nature, the representatives of geographical determinism failed to see the fundamental character of the differences between nature and society and the indirectness of the ties between them and sought to establish a direct connection between natural conditions and the life of human society or even between individual components of the natural environment and social phenomena.

One of the most outstanding and brilliant spokesmen of geographical determinism was Montesquieu (1689–1755).[6] The French Enlightener believed that the vast empires of Asia had arisen because of the vast plains there. Asia, he said, is carved up by mountains and seas into larger sections, and the rivers there are not significant obstacles to the movement of people, which is what promoted the formation of large states. Countries with fertile soil most often have a monarchic form of rule, whereas countries with infertile soils most often have a republican system. And Montesquieu attributed especially great importance to climate. "The peoples of hot countries are timorous, like old men, while the peoples of cold climates are brave, like youths. In hot countries, an excitation of passions is accompanied by an increase in crimes, and everyone tries to gain the upper hand over the other in everything that feeds these passions."[7] He ascribed the popularity of the Indian teaching about Nirvana to the influence of a hot climate that has a debilitating effect on people's mental capacities and produces a yearning for tranquillity. Montesquieu also ascribed the fact that slavery existed preponderantly in hot countries to the debilitation of southern peoples from the heat, and as a result they could work only under fear of punishment. He said that the power of climate was the strongest of all powers.

Although he linked the development of slavery to the debilitating influence of climate, Montesquieu wrote, not without a large dose of irony, that "there is probably not such a climate on earth under which labor could not be free."[8] He did not regard natural conditions as the only force determining social development. "Many things govern people: climate, religion, laws, principles of government, examples of the past, morals, customs; as a result of all this the general spirit of the people is formed."[9]

Despite their manifest untenability in the eyes of our contemporaries, Montesquieu's views had a progressive significance for their time. They ended in the conclusion that if a country's laws do not correspond to the natural environment and, consequently, to people's personalities, then these laws must be changed. To a large extent, Montesquieu was supported by Rousseau, who asserted that "the more one thinks about this principle, established by Montesquieu, the more one is convinced of its truth . . . ."[10]

The materialism of the eighteenth-century French philosophers extended only to the understanding of nature. They looked at history through "the eyes of idealists. To the extent that they dealt with the history of human societies, they tried to explain it through the *history of thought.* To them, the famous proposition of Anaxagoras that 'the intellect (*Nus*) rules the world' reduced to the proposition that human *reason rules history.*"[11]

The one-sided attribution of social phenomena to the direct influence of geographic conditions met opposition from a segment of the eighteenth-century French materialists. Some of them spoke of the scientific untenability of attempts to explain social life by the influence of the natural environment alone. Thus, Helvetius (1715–71) decisively rejected climate as a factor determining social phenomena and pointed to the possibility of reactionary conclusions from Montesquieu's geographic thought. Ridiculing geographical determinism, he wrote in one work: ". . . an obese Englishman who eats butter and beef and lives in a humid climate is definitely not more intelligent than an emaciated Spaniard who eats onions and garlic in a very dry climate";[12] in another, "If Italy was so rich in orators, it was by no means because the soil of Rome, as some academic pedants have asserted in their scholarly inanity, was more propitious for the creation of great orators than the soil of Lisbon or Constantinople. Rome simultaneously lost both its eloquence and its freedom, yet nothing happened to the soil, and the climate of Rome did not change under the emperors."[13] Diderot took a somewhat more cautious approach to solving the question of the significance of the geographic environment for social life. "We shall not ascribe too much importance to these causes, nor shall we reduce them to nil,"[14] he wrote, meaning man's environment.

One could cite a great many statements by scientists and philosophers that pointed to the scientific untenability of attributing social phenomena to the determining influence of natural conditions. For one more example, we turn to the great Russian thinker N. G. Chernyshevsky, who wrote: "To an Englishman, German or Frenchman, Italy is already south, and its climate is murderous to energy. Greek and Roman writers found, by contrast, that only Greece and Italy had a temperate climate that developed energy, while to the north, beyond the Danube and the Alps, the climate was already so inclement that civilized life could not be developed. What is south and what is north to each of us? This division simply depends, after all, on the latitude to which we ourselves have become accustomed."[15]

Geographical determinism had one of the strong elements inherent in materialism, including eighteenth-century French materialism, to wit: It was based on the idea of a unitary material world; it was monistic. "Man is the handiwork of nature, he exists in nature, he is subject to its laws, he cannot rid himself of it; he cannot escape from nature even in his mind."[16] "If by nature we shall mean an assemblage of dead substances, without any properties and utterly passive, then, of course, we shall have to seek outside of this nature the principle of its movement; but if by nature we shall mean that which it is in reality, to wit, a whole whose various parts have various properties, act according to these properties and are in continuous interaction with each other ..., then nothing will compel us to have recourse to the operation of supernatural forces in order to comprehend the formation of the things and phenomena we observe."[17] "Thus, if we are asked where matter came from, we shall answer that it always existed. If we are asked where the movement of matter came from, we shall answer that on the same grounds it has had to move since time immemorial, insofar as movement is a necessary result of its existence, its essence, and its original properties, such as dimension, weight, impenetrability, shape, and so forth."[18]

Furthermore, until the appearance of Marxism, criticism of geographical determinism lacked the most important attribute: it did not contrapose to this concept a new theory that would make it possible to penetrate more deeply into the mysteries of matter. Thus, it was primarily for this reason that over a long period geographers believed in geographical determinism or made mistakes based on it.

Here are a few examples supporting our opinion that most of the geographers of the past were adherents of geographical determinism in one form or another.

Thus, Alexander Humboldt (1769–1859) wrote: ". . . I have tried everywhere to show the perpetual influence of physical nature on the moral system and on the very fate of mankind."[19] Another major geographer of the same period, Carl Ritter (1779–1859), affirmed that England, "being in the center and surrounded on all sides by straits, on the strength of this became by herself the sovereign of the seas."[20] He also originated the widespread dictum that man is "the living mirror of nature."[21] The well-known Russian geographer, Academician K. M. Ber, asserted that "the fate of peoples is determined in advance and seemingly inevitably by the nature of the locality they occupy . . . ." "The physical properties of localities seemingly predetermine the fate of peoples and of all mankind . . . . The course of world history, of course, is determined more by external physical conditions."[22]

The concept of geographical determinism was also clearly manifested in the works of such essentially dissimilar scholars as Ratzel and Reclus. "The state as an institution is just as old as the family and society, from which it is distinguished primarily by its close relation to its country, which we may safely call a geographic property. Once a family chose a certain parcel of land to settle and cultivate and fenced it in against intrusion by foreigners, against attack by wild animals, and, finally even against floods from a nearby stream, there was in this case created that *unity of a people with a certain area of earth* which we call a state."[23] "What economist, what geographer or historian would deny the decisive influence of these geographic conditions on the course of events? In the quiet of one's study room, it is pleasant to succumb, as Nietzsche, Gobineau, or Driesmans do, to dreams of a superman and to affirm that our environment is in ourselves. This, if you like, is sublime, but it is absurd. No matter what Schiller sings about it, the continent of both Americas did not reveal itself to Columbus for the sake of fulfilling his dream."

"Of course, we are not saying that the environment is permanently immutable: much to the contrary, it is continually changing together with history, which is nothing but the evolution of the surrounding milieu, produced by the surrounding milieu itself. Today's tilled plain is no longer the blossoming steppe of former days; woodless mountainsides are no longer covered with mysterious thickets; the steppe no longer disgorges savage hordes; the swamps no longer serve as havens for enemies."[24]

"In any natural region, the contrasts of the soil, vegetation, and products of a country are accompanied by contrasts in the character and occupations of its population. The environment accounts for the characteristic differences that are observed in human society; it also explains why a certain low form of civilization can sustain itself through the centuries, while nearby farming nations living in conditions that are favorable for raising useful plants can make more or less rapid progress."[25]

It should be pointed out that geographical determinism, even in its most simplistic form (in which the development of social life was attributed to the direct and determining influence of the natural environment), still afforded an opportunity to form a correct understanding of many geographic phenomena in the world of nature, especially when, starting with Humboldt, it began to be combined on a wide scale with the comparative method. As soon as attempts were made to use "geographism" to explain social phenomena, it invariably led to unscientific, and ultimately to extremely reactionary, conclusions. By accepting the nature surrounding man as a sole and directly operating factor conditioning virtually the entire development of social life, geographers and philosophers switched from materialist positions to positions of historical idealism, since they held social development to be predetermined by the external conditions of the natural environment. Geographical determinism was one of the manifestations of metaphysics in pre-Marxian philosophy, with its inherent unhistorical treatment of human society and with its attempts to explain its development by some single eternal factor. The causal basis of all phenomena was frequently reduced to fatalism, which denied the significance of the active, purposeful operations of human society.

Geographical determinism as a philosophic basis for comprehending geographic phenomena should be distinguished from certain errors based on it that were made by many geographers who, however, mostly held positions of instinctive materialism and, in a number of cases, took steps in the direction of materialistic dialectics.

We shall not have a chance in this work to show the differences in the views of certain philosophers and geographers who either completely supported a platform of geographical determinism or made certain mistakes based on it.

This is a highly interesting subject for a special study. But we cannot fail to take note of the sweeping misidentification of virtually all past geographers as adherents of geographical determinism, despite the very substantive and fundamentally important differences in their views and approaches to the study of geographic objects and phenomena. For example, it would hardly be right to identify entirely as an adherent of geographical determinism the outstanding nineteenth-century Russian geographer, Lev Mechnikov (1838–88), as is frequently done. Mechnikov, who undoubtedly committed geographical-deterministic errors, nonetheless was able to understand the mutual character of the influences existing between society and nature and the occurrence of interactions between them. He took what was unquestionably a major step away from geographical determinism in the direction of materialist dialectics when he wrote: "We are far from *geographical fatalism*, which the theory of the influence of environment is frequently accused of espousing. In my opinion, the cause of the origins and the character of primordial institutions and their subsequent evolution are to be sought not in the environment itself but in the correlation between an environment and the capacity of the people inhabiting that environment to cooperate in solidarity. Thus, the historical value of a given geographical environment, even if one assumes that it remains physically unaltered under all circumstances, varies from historical epoch to historical epoch . . . ."[26] In attributing the unevenness of the territorial distribution of civilization to the geographic factor, Mechnikov contraposed his own, largely correct views to the reactionary concepts that derived the level of civilization from the racial features of the population.

It should also be emphasized that although Mechnikov failed to eschew geographical determinism altogether, he was infinitely far away from those reactionary interpretations of it that we sometimes get from bourgeois geographers. Although he advanced the geographic factor as the chief "motor" of history, Mechnikov always stressed that this factor exerted its influence not directly but in the interaction between society and nature and in the process of labor. "The river of nourishment compelled the population, under fear of inevitable death, to join efforts in common work and taught it solidarity, although in reality certain groups of the population may have hated each other. The river imposed on each individual member of society a certain part of the social task, the utility of which was recognized later, but at first was obscure to the great majority."[27]

Mechnikov's theory was of great scientific importance for its time. His book retains considerable scientific interest even in our day; one cannot overlook it when becoming familiar with the history of ideas in geography. Moreover, it is still valuable as a vivid and scientifically based work directed against the unscientific absurdities of the racists. One cannot help but regret, therefore, that the fate of the book, like the fate of Mechnikov himself, incidentally, was a tragic one.[28] A critical appraisal of Mechnikov's book was offered by Plekhanov: "Hence, although his book leaves no doubt whatsoever that the geographic environment influences man principally by means of the *economic relations that arise under its effect,* he gives very little illumination in the book to the economic side of the matter."[29]

A large role in the affirmation of geographical determinism was played by Henry Buckle (1821–62). "If we undertake to consider what physical factors have the most powerful influence on humankind, we shall find that they may be placed under four major classes, to wit: climate, food, soil, and the general type of nature."[30]

Production and distribution are entirely conditioned by these four conditions, just as the basic differences between the European and non-European peoples are, affirmed Buckle. As we see, he expressed the very valid idea that the geographic environment influenced the development of society through production; however, he did not develop this idea.

The observation by Helvetius that it was possible to derive not only progressive but also reactionary conclusions from the theory of geographical determinism was entirely confirmed in the case of Buckle, who wrote: "Energy and regularity in labor itself depend absolutely on the influence of climate." This implied a repudiation of explanations of social phenomena based on divine action, but the materialistic explanation of the process of historical development of society was elevated to a natural law that supposedly affirmed the eternal order of capitalist exploitation and colonial oppression. Thus, geographical determinism, which served as a weapon against theology, was gradually turned into an ideological justification of capitalist slavery.

Nevertheless, the majority of geographers of the past avoided carrying over natural laws into the sphere of social relations; and in the instances in which it was done, they still did not go so far as to draw such far-reaching reactionary conclusions, as did the representatives of contemporary geo-graphical determinism. On the other hand, an understanding of the definite unity of the material world and the intercausal-ity in the development of nature and society gave geographers a theoretical basis for correct explanations of many geo-graphic phenomena and made it possible to compose generalizing regional-geographical works that presented a complete picture of countries and regions and not one torn up into individual elements.

Errors based on geographical determinism did not prevent many materialist geographers of the past from writing valu-able, scientific geographical monographs and performing im-portant research on the earth's outer envelope. In speaking of regional-geographical works, the first that should be identified is the monograph, *A New Universal Geography: The Earth and Its Inhabitants*, written by the French geographer and revolutionary, Élisée Reclus (1830–1905). "The regional-geographical theory of Reclus is still striking in its breadth of outlook and its capacity to weave the data of physical, histori-cal, economic and political geography, demography, ethnog-raphy, and observations of culture and everyday life into an overall geographical picture."[31] Incidentally, Reclus was the first to use the term "geographic environment," meaning by it the conditions of social development that surround man. It is characteristic that Reclus accurately defined the essence of the geographic environment as a combination not only of natural but also of social elements, which he called "dynamic." He wrote: "And so, the entire environment re-solves into an infinite number of individual elements: some of them belong to external nature, and they are usually identified as the 'outer environment' in the narrow sense of the word [i.e., "natural environment" in modern terms — V. A.]; others belong to a different order, since they stem from the very course of development of human societies and are formed by increasing consistently to infinity, multiplying and creating a complex variety of phenomena in action.

"This second, 'dynamic' environment [i.e., "social envi-ronment" in modern terms — V. A.], by adding itself to the influence of the first, 'static' environment, forms a sum of influences in which it is difficult and often even impossible to determine what forces predominate."[32]

Reclus largely eschewed direct geographical determinism. He saw the operation of internal laws of the development of human history and understood the historical character of the influence of the geographic environment on the life of society. "And so, the history of mankind, both in its entirety and in its parts, may be explained only by the total influence of external conditions and of complex internal strivings through the centuries. However, in order to gain a better understanding of the evolution that is taking place, it is imperative also to consider in what measure the external conditions themselves are changing and in what measure, consequently, their effect is changing in the overall evolution.

For example, a mountain range from which colossal glaciers once descended into the neighboring valleys preventing anyone from climbing its steep slopes, later lost its significance as such an obstacle to communication between neighboring peoples once the glaciers receded and only its crest remained covered with snow. In precisely the same way, a river that was a formidable obstacle to tribes unfamiliar with navigation could later be made into an important navigational artery and take on enormous importance in the life of the population along its banks when this population learned to navigate boats and ships."[33]

It is also important for us to stress that Reclus wrote his regional-geographical monograph with great love for man and his work. It is devoid of racism and is pointed entirely toward the future. Reclus skillfully and truthfully depicted man's struggle against nature and the changes occurring in nature and society as a result of that struggle. Reclus was not a Marxist. And what is more, like Mechnikov, he opposed Marxism and adhered to essentially anarchic ideas. But his world geography, which is filled with abundant factual material, nevertheless depicts many aspects of the interaction between nature and society with surprising accuracy.

It is interesting in passing to recall the message from Reclus to his Russian readers, in which he speaks of the Russian people with great warmth and prophetically predicts for them a great future.

"And you Russians, what part will you take in this broad movement that is carrying you to the entrance into a new world? . . . What will they think of you? What great talents will history generously endow you with?

"We can answer this in advance: Everyone will have to recognize as your principal attribute the fact that you were the most hospitable, the most fraternal people.

"A nation that has embraced an endless plain that joins with other plains via magnificent paths, you possess at the same time both the qualities of a firmly settled farmer who loves the earth and tills it with tenderness and the free nature of a nomad who feels at home everywhere, be it in the north, in the icy tundra of the White Sea, or in the south, among the vineyards and burning limestones of the Crimea . . . . You will be welcome guests everywhere and will receive everyone in your own country as friends; no national group will contribute as much as yours to the birth of the nation of the future, which will descend from all races and will speak all languages. You will be the principal actors in the creation of a truly human civilization, based on freedom and law."[34]

The representatives of the widely known French school, "Géographie Humaine," also created a large regional-geographical study of the world, "Géographie Universelle," which unquestionably is one of the outstanding achievements of world regional geography.[35]

Major regional-geographical studies were also written by Russian geographers, including the outstanding multivolume *Russia*, edited by P. P. Semenov-Tyan-Shansky. In general, the determinist geographers of the nineteenth century made a major contribution to geography.

S. M. Solovyov, A. P. Shchapov, and V. O. Klyuchevsky, the major representatives of Russian prerevolutionary historical science, attributed many peculiarities in the historical development of Russia to the direct influence of the natural environment.[36] Enormous importance was also attached to the geographic factor in the history of mankind by N. V. Gogol, who left us a number of interesting statements about geography. For example: "But above all one must cast one's eye upon the geographical situation of this country, which should certainly precede everything else, because the way of life and even the character of a people depend on the type of land. Much in history is decided by geography."[37] Of course, it is easy now to criticize geographical determinism and its proponents, but if one recalls that such determinist explanations were contraposed to indeterminist concepts that held the will of God and the tsar to be the chief factors of historical development, then one cannot fail to recognize the positive value of geographical determinism in the development of Russian historical science. Moreover, the depiction of the historical process against a geographical backdrop and the establishment of actual interaction between society and was in itself a major scientific achievement.

Incidentally, the works of Soviet historians sometimes suffer from inadequate attention to the influence of the geographic environment on social development. Perhaps the fear of being accused of geographical determinism leads them on occasion to the other extreme, to the dissociation of the historical process from the concrete geographic conditions in which this process is taking place.

Even in the area of natural geographic phenomena, we still frequently encounter inattention to the interconnections that exist between them and an insufficiently deterministic approach to their study. The interconnections between the elements of the geographic environment within the natural complex are not always taken into account in specific geographical research, and geographers often do not consider it possible "to submit the earth's vastness to a single view," that is, in effect, they repudiate geography.

B. B. Polynov made an astute observation in this regard. "There are instances in which the truth does not encounter objections and apparently receives general recognition, but at the same time seems to remain outside the frame of consciousness, and at every turn things are done that contradict it. This is precisely the situation in our country concerning the truth about the interrelationship between occurrences and phenomena in nature."[38]

As concrete knowledge about the earth and its individual parts was accumulated, new possibilities arose for generalizing works. Yet here it would be wrong to believe that the transition to theoretical generalizations became possible only after the process of describing continents and oceans was completed. Such generalizations had been carried out previously in ancient geography, but the quantity and quality of the accumulated factual material are of extremely great importance. It is therefore impermissible in geography to contrapose the accumulation of factual material to its generalization, or description to analysis and synthesis. Yet, in our country, one sometimes comes across such contrapositions, which lead to the nihilistic allegation that all of geography was unscientific until the end of the nineteenth century. In no way can one agree with the contention that "it is precisely for this reason that one can place the origin of scientific geography at the turn of the twentieth century."[39]

The development of science and its level, as a rule, are consistent with the overall level of development of productive forces that has a been achieved by a given people at a given time. It is therefore totally wrong to evaluate the science of an epoch in terms of another, later epoch. Ancient geography was quite scientific for its time, just as medieval geography was scientific for a certain period of feudalism. It is indisputable that the geography of Eratosthenes was no less scientific for its time than the geography of Humboldt for *its* time. If one goes along with I. M. Zabelin, then it is not hard to predict that twentieth-century geography will also be declared unscientific, for example, in the twenty-second or twenty-third century, if the geographers of that time also include adherents of a nihilistic attitude toward history. The contraposition itself of scientific to unscientific geography cannot be considered at all scientific. Geography was not created anew at the turn of the twentieth century. That period was merely a new stage in the development of the same geography that had already existed for several millenia. A process of intensive differentiation of geography occurred in the nineteenth century, something that was of great positive value in the development of geography; but the end of the nineteenth and the beginning of the twentieth centuries were characterized by an extreme lack of synthesis, which made it possible to advance theories denying the unity of geography.

Furthermore, the actual status of geography always disproved a nonhistorical approach to the evaluation of its progress. If one speaks of eighteenth-century geography, by that time general geography, or earth science, had been given a sound scientific groundwork, since it became possible, because of the development of means of astronomical observation, to accurately determine the geographical location of places (latitudes and longitudes) and to perform degree measurements by modern methods, that is, approach the solution of the question of the form of the earth. From that time on, from the "age of measurements," as Peschel called that period in the history of geography, it became possible on a higher scientific basis to write works characterizing the earth's outer envelope, both as a whole and in its individual parts.

CHAPTER THREE

Concerning the Influence of
the Philosophies of Kant and Hegel
on Geography. Alexander Humboldt and
Carl Ritter. Elemental Materialism and
Dialectical Idealism in Geography.
Origins of Geopolitics. Hettner's Theory.

Thus, geographical determinism was associated, to a considerable extent, with the materialist school in bourgeois philosophy, which reached its peak in France. However the development of theoretical concepts in geography also proceeded under the influence of the idealist school in bourgeois philosophy, and that school was most developed in Germany. Geography, above all German geography, was especially influenced by the idealist philosophers Kant (1724–1804) and Hegel (1770–1831).

The world outlook and scientific activities of Kant represent one of the most complicated and contradictory phenomena in the history of ideas. Kant provided a number of new, basically materialistic natural-science theories that advanced science. It is common knowledge that he created the materialistic hypothesis of the origin of the solar system, which dealt a blow to metaphysical views of the world, thanks to a historical approach to the problems of cosmogony. From 1757 to 1797 Kant taught physical geography at Königsberg University. His lectures were a significant achievement in geography for their time. The generalization of a large amount of factual material was consistent, on the whole, with the ideas of Varenius, which had remained the most progressive throughout the eighteenth century: Kant viewed general and regional geography as a unity, as two interrelated divisions of one science. This was a step forward by comparison with the still widespread concepts that separated the study of the earth as a whole from the study of its individual parts.

But at the same time, in philosophy, Kant sought to reconcile the irreconcilable, to establish a compromise between materialism and idealism and combine in one philosophic doctrine two opposite, mutually exclusive bases. In contending that our theories imply something external to us, a "thing in itself," Kant gave the materialistic world view its due. In his concrete scientific investigations of nature, which he regards as a "thing in itself" in his philosophical system, Kant is a materialist; but when "he declares this thing in itself to be unintelligible, transcendental, and otherworldly, Kant is an idealist."[1]

Kant acknowledged the existence of causal ties between society and nature. Peoples, in his opinion, are separated from each other by natural boundaries (mountain ranges, major rivers, etc.). He regarded the violation of such borders as a violation of a certain law-governed equilibrium, a violation that inevitably leads to bloody wars. Kant also linked production activity to the natural environment and resources. On the basis of geographical determinism, Kant also defined the object of study of political geography: "the situation of these countries themselves, the works of man in them, their morals, handicrafts, trade, and population."[2]

One can obviously conclude that Kant viewed the influence of the natural environment on society primarily as the influence of the geographic conditions of social life that promote production activity. This sort of approach to evaluating the influence of nature on society was a step forward, since before Kant the influence of nature on society was viewed chiefly as a physiological influence.

Kant's natural-science views contained quite a number of ideas that were progressive for their time; and their influence on geography, as well as the influence of the geographical works themselves, was initially positive and strengthened the determinist perception of the world in geography.

But the inconsistency of Kant's philosophy made its influence on science, geography in particular, contradictory. In addition to affirming the determinist world view and the great positive value of his firsthand scientific activity in cosmogony and geography, Kant gave grounds for numerous idealistic distortions in understanding the nature and basic tasks of geography.

Assuming the impossibility of knowing the objectively existing world, Kant affirmed that time and space were subjective forms of perception, thereby separating them from matter.

"Time is not something objective or real, it is neither a substance, nor an accident, nor a relation, but a subjective condition that is necessary, by the nature of the human spirit, for the coordination of everything sensible according to a certain law and is pure contemplation."[3] *"Space is not something objective* or real, it is neither a substance, nor an accident, nor a relation, but it is *subjective* and ideal: it is a schema that seems to have originated from the nature of the spirit according to a constant law in order to coordinate everything that is perceived from outside."[4]

The dissociation of the categories of time and space from matter led Kant to create a classification of sciences that did not reflect the material reality of the objects they dealt with. According to this classification, geography was assigned the role of *describing* phenomena that coexist simultaneously in space. History was also relegated to the purely descriptive sciences, but its task was reduced to a description of events that occur one after another in time. "History is a narrative, whereas geography is a description . . . ."

Thus, geography was turned into a specific science of the components of space or of the distribution in it of objects and phenomena.

In assigning geography the role of merely describing objects and phenomena in space, Kant rejected the possibility of examining their development, since that process, according to his concepts, was under the purview of another science — history. In separating space from matter, Kant attempted to found a special spatial science, which was a logical outgrowth of his idealistic world view.

The elements of subjective idealism that were contained in Kant's philosophy led very clearly in the direction of limiting reason and strengthening faith, which greatly influenced geography. This influence was expressed in the denial of a universal causal connection between the phenomena of the material world and in the absolute contraposition of human society to the rest of nature, that is, it led to indeterminism in geography or, in other words, to geographical determinism. The Kantian conception of geography implied that knowledge of the geographic environment as a whole was impossible, since the latter was in constant development. Geography was identified as a science composed of individual sciences not related organically to one another, each of which was capable of dealing only with separate, particular phenomena.

Thus Kant realized that descriptive geography could not explain the essence of phenomena of the material world that were under study, that the history of their development had to be studied for this purpose. But for him history differed sharply from geography. "The description of nature (i.e., the state of nature at the present time) is far from sufficient to indicate the basis for explaining the whole variety of its changes. It must be resolved, despite all of the very justified hostility to boldly proposed opinions, to create a *history* of nature that would be a separate science and would gradually prove capable, needless to say, of progressing from simple opinions to well-founded knowledge."[5]

Geography, according to Kant, is incapable of ascertaining the conditions of development of society primarily because these conditions are the result of development, and it is impossible to understand the result without discovering the causes. Finally, Kant widened the gap in geography between the study of nature and the study of society, because nature, in his conception, was subordinated to dead laws that reason did not understand. The sciences of nature, therefore, cannot understand development and have nothing in common with the social sciences, whose objects of study are particular spiritual substances.

Kant rejected determinism in the area of social life. He held any dependence of practical reason or human will on external causes to be unscientific heteronomy. According to Kant, people live and act primarily on the basis of purely moral laws that are independent of material or sensual motivations. Although he recognized determinism in natural science and even saw the causal character of the ties between society and nature, he patently contradicted his own propositions by contraposing human society, as a special spiritual sphere, to the rest of nature. In this way the groundwork was laid for the indeterministic conception of geography, in which the two conclusions below, derived from Kant's philosophy by his followers, are of the greatest importance: (1) Space and time were divorced from matter, and consequently from practice, since practice cannot exist without matter. As a logical corollary of this, geography was assigned the role of dealing only with territorial relations. (2) The treatment of geography as a purely descriptive science. Geography describes everything on the earth's surface and describes the location of everything that exists on earth.

The contradictory character of Kant's philosophy later made possible its wide use for indeterministic distortions. Kant's followers, who·developed the idealistic aspects of his doctrine, contributed a great deal of idealistic confusion to geography by strengthening its indeterministic tendencies. It should also be noted that the Neo-Kantians made use of those more reactionary aspects of Kant's philosophy associated with the influence exerted on him by the philosophy of David Hume (1711–76), which represented an attempt to substantiate a patently indeterministic view of the world.[6] The elements of subjective idealism in Kant's philosophy were used and developed on an especially wide scale by representatives of the Baden (Freiburg) school of Neo-Kantians, who advanced the idea of a total cleavage between the natural and social sciences.

The Baden Neo-Kantians proposed that sciences be classified not by their objects of study but by their point of view, by their purpose of inquiry. One of the school's most prominent representatives was Rickert (1863–1936), who made rather wide use of geographical material for his philosophic constructs.

Rickert divided all sciences into generalizing (which comprised the natural sciences) and individualizing sciences (which comprised the social sciences).[7]

An indeterminist partition was erected between nature and society, they were declared to be unrelated; and no general laws were recognized as operative in the sphere of social relations. Hence the social sciences could deal only with solitary, individual facts.[8]

In connection with this division of sciences, two geographies were identified. One of them, which dealt with the earth's surface as a whole, came under the generalizing (i.e., natural) sciences and was combined with geology; the other, which dealt with the earth's surface in terms of the development of human culture, was declared an individualizing (i.e., social) science.

Such Neo-Kantian views also acquired currency among Russian prerevolutionary geographers. "A direct consequence of idealistic dualism in Russian prerevolutionary geography were the views that the very essence of the external world — nature — is unintelligible and that there is no such thing as objectively existing territorial combinations of productive forces that develop according to certain laws .... In short, the incompatibility of the two divisions of geography [i.e., physical and economic — V. A.] within the framework of a single science was established by bourgeois Russian geographers on the basis of idealistic philosophic positions, which were absolutely unacceptable to Marxism."[9]

The views of the Baden school's representatives that knowledge is restricted to the world of phenomena (essence was declared unintelligible) and that the natural and social sciences are distinctly separate are still current in other countries, especially the United States and the Federal Republic of Germany, although they are not dominant there.

For example, in our time the German sociologist Theimer is attempting to contrapose to Marxism his own (essentially Rickertian) Neo-Kantian theories. According to him, the best solution of the question of the relationship between the natural and social sciences is that of the Baden Kantians, among whom he rates Weber especially highly. By denying the existence of any general laws governing both nature and society and denying an interconnection between them, Theimer completely contraposes the social sciences to the natural. Social phenomena in general, as a result of their great diversity, are deprived by Theimer of any general patterns, which are permitted to exist only in the world of nature. Because of the absence of general laws governing the development of society, sciences dealing with it are not capable of elaborating general concepts and can deal only with individual phenomena that are of interest to us in each specific instance.

Although Theimer recognizes the materiality of nature but denies the dialectics of it, all the natural sciences are based on mechanical materialism, which allows particulars to be known but eliminates the possibility of knowing nature as a whole; this should confirm Kant's proposition on the unintelligibility of essence. This kind of mechanical materialism is capable, according to Theimer, of ensuring comprehension of some common elements and mechanical aggregates of nature and makes it possible to generalize phenomena of nature, especially inert nature. In the social sciences, meanwhile, the will of individual personalities and their ideals are declared to be the determining factors of social life. Such Neo-Kantian concepts leave no room for geography as a science.

In breaking up the unity of the material world, Rickert and his followers made use of Kant's gnosiology to disprove the theory of the objectivity of the laws of the historical process, seeking to prove the impossibility of scientific knowledge and scientific prognosis in the field of history.

The indeterminist philosophy, which denied the unity of the material world and broke it down into unrelated categories, was the basis for the so-called office statistics [financial or administrative information and data — Eds.] that developed in Germany in the eighteenth century, a development that was aided in no small measure by the needs of the bureaucratic semifeudal apparatus of German states.

These office statistics included geographical information in the form of completely unrelated data that candidates for bureaucratic positions were supposed to learn by rote. This was a jumble of various unrelated data. Commercial geography was the same kind of jumble of data on trade items and on the technology of some industries.

Office statistics and commercial geography existed for a very long time in somewhat altered form, making up the basis for the statistical trend in economic geography, which has survived, as we know, right up to the present.

N. N. Baranskiy made a keen observation about this trend, noting that the development of scientific thought proceeded "not within economic geography itself but in disciplines associated with it . . . ."[10]

Hegel's philosophy, which, as we know, was the highest form of objective idealism, had a different influence on geography. Hegel considered problems of geography in connection with his general-philosophic and natural-philosophic concepts.

The principal contradiction of the Hegelian theory is that between the dialectical method, which views all objects, phenomena, and processes in continual development, and the idealistic metaphysical system, which restricts this development to certain previously formulated boundaries in the form of a system of categories elaborated by Hegel in his *The Science of Logic*. As an idealist, Hegel ascribed development to the single realm of the spirit, which revealed itself in various areas of human activity and especially in the sphere of ideology (religion, arts, philosophy, etc.). On the other hand, matter, according to Hegel, is something inert and passive. In contrast to the spirit, it is not capable of self-development. In effect, nature is not capable of it, either. In Engels's definition, "for Hegel, nature, as the simple 'alienation' of an idea, is not capable of development in time; it can only develop its manifold features in space, and thus, condemned to eternal repetition of the same processes, it exhibits all of its levels of development simultaneously and side by side."[11]

In his *Philosophy of History,* Hegel developed a theory postulating the existence of historical and nonhistorical peoples, which is still used in various modifications as a philosophic basis for racist pseudodoctrines. His theory linked historical peoples to certain natural conditions that were matched with certain territories; "... the special principle, embodied in every world-historical people, is, at the same time embodied therein as a national characteristic."[12] Therefore, geography was considered a science that dealt with nature as the basis of the historical development of human society.

The flaws in Hegelian methodology led to the point where the study of nature was deemed possible only in the form of the description of space, and this was declared the task of geography. Notwithstanding this metaphysical tenet, the strength of the principles of dialectics that Hegel formulated had an influence on some geographers, which was reflected favorably in the development of geography. These dialectical principles were: (1) the approach to phenomena in terms of their origin, development, and destruction; (2) establishment of the cause of development as a consequence of the contradictions inherent in every object; (3) solution (i.e., elimination) of the contradictions not only by means of gradual, quantitative change but by means of transition into a new quality.

The dialectical proposition on analysis and synthesis developed later by Marx and Engels was of especially great importance for geography. Hegel held that the dialectical method "in all of its motions is at once analytic and synthetic."[13] This proposition remains today one of the most important for a correct understanding of the essence of geography. Hegel's philosophy thus contained dialectical propositions that could be directed against the positivist view of geography.

By striking a blow at the metaphysical world view, Hegel made a noticeable stride forward in the determinist view of the geographic environment as a causally related unity of various elements. In very general form he called attention to the indirect character of the natural environment's influence on the life of society and established the existence of a causal relation between nature and society. "It is not our concern to become acquainted with the land as an external locale,"

Hegel writes in his *Philosophy of History,* "but with the natural type of the locality, as intimately connected with the type and character of the people who are the offspring of this soil. This character is nothing more or less than the mode and form in which nations appear in world history and their place and position in it. The significance of nature should neither be exaggerated nor belittled ...."[14] Although Hegel never did succeed in solving the question of the geographic environment's role in the life of society, it would be wrong to treat the influence of his philosophy on the development of theoretical concepts in geography only as something negative and occasionally encountered. True, in his efforts to solve specific questions in geography, Hegel accomplished nothing new; but, as has been mentioned, Hegel's dialectic, taken up by some geographers even in its idealistic form, determined a new, more correct approach to the objects of study of the geographical sciences.

This new approach was carried out in different ways. Some geographers, especially Russian, who adhered to positions of instinctive materialism, tried to apply dialectics on a materialist basis, which yielded highly favorable results in the study of the physical and biological elements of the landscape envelope. As we shall attempt to show later, the one who came closest to positions of dialectical materialism was the great Russian scholar V. V. Dokuchayev. Other geographers, who held to their idealistic positions, used the method of idealistic dialectics.

It should be stressed that the boundary between the materialist and idealist schools in geography has always been quite clear-cut. To show the basic differences in the theoretical concepts and concrete research between these two principal schools, it suffices, at least briefly, to familiarize ourselves with the theoretical concepts of Humboldt and Ritter. The works of these outstanding German geographers of the first half of the nineteenth century probably displayed in the most vivid and graphic way the two principal schools in the development of geography. Moreover, the ideas of Humboldt and Ritter had a noticeable influence on many geographers after them, which compels us to devote a little more attention to their theoretical views.

First of all, let us note the features that were common to the concepts of Humboldt and Ritter. These common elements consisted in their *recognition of the unity of geography*. Both recognized that the division of geography into the general and particular (or into earth science and regional geography) was a division into parts of *one general science*. Both also recognized that geography should provide integrated pictures of man's environment, including man himself, with the results of his work and some distinctive characteristics of social life. Both, therefore, recognized that geography had one general object of study, despite the differentiation of geographical research. Finally, both were somewhat influenced by Kant's philosophy: they used elements of dialectics, considering it necessary and possible to study geographic phenomena in the process of their development.

These common features in the theoretical views of Humboldt and Ritter are very substantial, and it would be wrong to just contrapose these two scholars to each other. But it would be an even greater mistake to regard them as geographers who stood *entirely* on a *common* theoretical foundation and only complemented one another simply because in one instance investigative thought was directed toward studying nature (Humboldt) and in the other, toward studying the geography of society (Ritter). The rather widespread conception that the views of Humboldt and Ritter made up a mutually complementary unit is, in our opinion, totally incorrect. It arose because of the overestimation of the common features in their views of geography and the disregard of the highly substantive differences between them.[15]

Humboldt's works were largely based on generalizations of the large amount of empirical material that had been collected up to his time. Of great value in the accumulation of this material were the expeditions of the eighteenth century, in which discoveries and cartographic location of newly discovered lands were accompanied by systematic scientific descriptions of a regional-geographical character. "Geography was elevated to the level of a science by the determination of the shape of the earth and the numerous voyages, which only now began to be undertaken with benefits for science."[16] Humboldt himself was an inquisitive traveler and naturalist who knew how to combine concrete expeditionary inquiry into nature with broad generalizations and profound inferences derived on the basis of many years of personal participation in the concrete study of the earth's landscape envelope.[17] From this alone it is evident that his path in science was fundamentally different from that of the armchair scholar Ritter. If one looks at the scientific results of Humboldt's work, one is still struck by their extraordinary fruitfulness.

In his views of geography, in his approach to the study of nature, and in his major generalizing works, Humboldt invariably based himself on natural-historical, instinctive materialism. In contrast to Ritter, he viewed the unity of the world as a material unity moved by internal forces without any divine foundation. "My great incentive has always been a desire to grasp the phenomena of the external world in their overall association, and nature as an entity that is moved and animated by internal forces."[18] He paid particular attention to the study of the structure of the earth's geographic envelope, pointing to its complex, synthetic character. He put forth highly fruitful ideas about the nature and essence of geography as a science. Of special interest in this respect is the methodological chapter, "Introduction to the Physical Description of the Earth," in his *The Cosmos,* where Humboldt, from a materialist position not lacking in dialectics, polemicizes, in effect, with the Kantian concept of geography: he comes out against the dissociation of space from time and against the Kantian classification of sciences, asserting that the description of anything that exists cannot be separated from its history. "But what exists in the concept of nature cannot be absolutely separated from its activity: because not only the organic is an unceasing activity and process, the whole life of the earth, at every stage of its existence, points to previously experienced changes. Thus, the strata of rock that lie one over the other and constitute the larger part of the earth's outer crust represent vestiges of creations that have almost completely disappeared .... These strata at once reveal to the observer the fauna and flora of different epochs that have collected in one place. In this sense it is impossible to separate completely the description of nature from the history of nature. A geognost cannot understand the present without the past. Existence in its extent and inner essence can be fully learned only as something that *has been made.*"[19]

In developing the ideas of Varenius, Humboldt did much toward a correct explanation of the unity of general and regional geography and — something especially important — opposed the mechanical view of nature as a simple sum of particulars, thereby striking a blow at the positivism that was widespread among geographers who had gone off completely into particular specialized research. With all of his characteristic thoroughness, Humboldt came out against the allegations that, as the individual subfields of geography (individual geographical sciences) develop, geography as a whole disappears, since there are individual objects of study for the individual geographical sciences, but no general object of study for geography as a whole. On this point he wrote: "Upon reasonable contemplation, nature is a unity in a multitude, a combination of the diverse in form and composition, it is a concept of a totality of natural phenomena and natural forces as a living whole. The chief purpose of a rational study of nature is to recognize the unity in the diversity and to grasp in the particulars everything that has been passed on to us by the discoveries of preceding centuries and of the present time, but in such a way, by checking details as to be able to choose between them and not become a victim of their mass ...."[20] "The deeper you penetrate into the essence of natural forces, the more you fathom the relation among phenomena. At the outset of human learning, all phenomena, being considered superficially, seem to stand separately and to resist any rapprochement; repeated observations and reflection bring them closer to each other and show their interdependence, and by this means there appears a great opportunity to simplify the exposition of general ideas and to make it more concise."[21]

One can find in Humboldt even more definite statements against the disjointedness of geography. Thus, also in *The Cosmos*, he writes: "Specialized descriptions of lands are, without question, the most necessary material of general physical geography; but the most painstaking codification of these accounts of various lands will present just as meager a descriptive picture of nature in its totality as the mere computation of all the flora in the world would represent the geography of plants."[22] These quotations from Humboldt, to a certain degree, have not lost their significance today.

Humboldt's work is devoid of mysticism and idealistic fantasies. His works are based on real facts and on the generalization of experience. Only real bodies, processes, and phenomena comprised the range of questions he considered. He saw the interconnections between natural components not only as interrelations in space. He contended correctly that spatial interrelations are impossible without the interaction of individual bodies and forces in the material world, just as the existence of material bodies without space is impossible. The clear understanding that the existence of nonspatial matter and nonmaterial space is impossible was one of the characteristic and strongest features in all of Humboldt's works. He saw the unity of the material world and in this respect, to a substantial degree, shared the viewpoint of the eighteenth-century French materialists. But Humboldt's methodology contained many more elements of dialectics than the philosophy of not only the French Encyclopedists but also of Ludwig Feuerbach, and this noticeably distinguished Humboldt among the other scholars of the first half of the nineteenth century.

In contrast to the idealist geographers, Humboldt affirmed that geography had its own actual material object of study, which should be dealt with in time, space, and interaction between its constituent elements. He underscored most vigorously the complex, synthetic character of geography's object of study, he stressed its unity and hence the specificity of geography as a science. "The distribution of organic types according to the latitude and altitude of places and to climates, the geography of plants and animals, is just as different from descriptive botany and zoology as the geological investigation of the earth is different from mineralogy. That is why the physics of the universe should not be mixed up with the so-called *encyclopedia* of natural sciences .... In the theory of the universe, the particular will be considered only in its relation to the whole, as a part of world phenomena."[23] Humboldt possessed a deep understanding of the complex and varied mutual causality that existed between the individual components of the natural environment. "The word *climate* signifies primarily, without a doubt, a specific property of the atmosphere, but this property depends on the unceasing *interaction* of the sea — which is deeply furrowed everywhere by currents of various temperatures and emits heat rays — and the land — which is manifoldly dismembered, elevated and colored, bare or covered with forests and grass."[24] It is perhaps difficult to find in any other geographer's work a more vivid example of the essence of one of the particular objects of geographic study (climate) in its causal and inextricable relation with other elements, than in the above quotation.

Taking a broad geographical approach to the phenomena of nature, Humboldt tried to prove his idea that geography is not a sum of information but an autonomous (or, as he wrote, distinctive) science. By discerning and formulating the patterns of development of the natural environment as a synthetic combination of its constituent elements, Humboldt treated geography as a natural science. His major work, *The Cosmos*, was devoted to a theoretical substantiation of physical geography.

Humboldt's works vividly embodied the best features of the stage of development in geography that Engels called "comparative physical geography." As previously mentioned, Humboldt generalized an enormous amount of factual material accumulated by geography before the beginning of the nineteenth century. This generalization became possible, thanks to the wide-scale use of the comparative method, with the aid of which he established a number of highly important geographic patterns, such as the law of the altitudinal zonality of climates, the law of the variation of the altitude of the snow line in mountains in accordance with geographical latitude and the character of the climate, and so forth. Humboldt did not confine himself solely to the comparative method, but succeeded in combining it with an integrated approach to objects of study. This afforded an opportunity to take a correct approach to a thoroughgoing study of nature and to penetrate deeper into the essence of individual components of the natural environment, which is what made his works the basis for the development of many branches of physical geography.

Even now we cannot disagree with the assessment of Humboldt that was made by one of the most outstanding representatives of Russian classical geography, D. N. Anuchin. "If one could name someone who should always remain in a prominent place in the history of earth science, then it would unquestionably be Alexander Von Humboldt."[25]

Although he recognized the unity of the material world and attempted in his works to embrace nature in all of its aspects, Aristotelian style, Humboldt at the same time failed to arrive at a correct understanding of geographic phenomena of a social character and to explain scientifically the essence of the interrelations between nature and society. He reasoned in the vein of geographical determinism. In perusing Humboldt's works, one can easily discover that, though he recognized the unity of the material world, he, in fact, studied almost exclusively the natural part of this unity and depersonalized nature. The unity of geography, according to Humboldt, therefore, meant the unity of physical (including biological) geography. In those instances in which Humboldt did touch on social phenomena, he usually limited himself to a description and the statement of instances of the purely external dependence of people's activity on natural conditions. His profound theoretical propositions, which were far ahead of their time, pertained to comprehension of the natural complex of the earth's landscape envelope. He viewed geography, as a whole, primarily as physical geography, as "the physics of the terrestrial sphere," which his *The Cosmos*, we repeat, was intended to substantiate. Humboldt's materialism was marred by inconsistency once it came to attempts to explain human society's qualitative differences from the rest of nature. It was from Humboldt's time or, more accurately, after him, that the departure into pure physical geography began to intensify among a considerable segment of geographers, often without any attempts to establish ties between natural and social phenomena. I. P. Gerasimov says about that period: "In the study of particular concrete relationships, especially in the sphere of natural phenomena, geography has long since achieved many outstanding results. However, the historical and genetic essence of the highly important general relationships between the natural and social phenomena described during the geographic study of various regions and countries was for a long time too difficult for consistent scientific interpretation. This area of scientific knowledge consisted chiefly of isolated correct guesses by outstanding thinkers of the past. But in most instances the general theory of geography was dominated for many centuries by false, idealistic concepts that to this day still feed the reactionary, unscientific views of some foreign geographers."[26] Thus, it is perfectly obvious in his statement that Gerasimov has in mind a longer time span than from Humboldt to today. However, if one speaks of outstanding results in the study of individual components of the natural environment with the simultaneous difficulty of ascertaining the essence of the relationships between natural and social phenomena, this situation, of course, occurred most strikingly after Humboldt and continues to manifest itself today.

Ritter's theoretical views were a mixture of idealistic propositions adopted from the philosophies of Kant and Hegel. In addition, he was greatly influenced by the works of the philosopher J. Herder (1744–1803), which were not lacking in materialistic tendencies. More specifically, Ritter's understanding of the unity of the world is tied to some of Herder's views that were marred by great contradictions and inconsistencies. Herder was closer to a materialist world outlook than Kant. He was an adherent of the natural origin of living organisms, including man, in which he anticipated Darwin. Herder was also a great believer in the ideas of geographical determinism. For example, he attached great importance to differentiated natural conditions in the development of human society. "The heterogeneity of people, like that of all earthly creations, has its cause in the heterogeneity of localities."[27] Herder criticized Kant from an essentially naturalistic and materialistic standpoint.[28] With reference to the influence of the natural environment, above all relief, on social life, Herder observed the unity between all the forms of the material world and observed the interrelation and interdependence between society and the rest of nature; but he inferred the unity of the world not from a material foundation but from a universal spirit that predetermines the destinies of peoples. Here Herder was a captive of idealistic concepts.

Geography, according to Herder, is the basis of history, and history is geography set in motion; geography deals with static space (as the arena of history), whereas history deals with developments occurring in time one after another. Space and time thus were also dissociated from matter by Herder, just as they were in the philosophy of Kant.

Ritter adapted his systematization of geographical materials under investigation, to theological schemes, using geography to prove both the divine origin of the earth and man and the fatal predestinationism of the development of society and nature. Ritter linked the interrelations that actually existed in nature and the relationships between it and society to religious ideas of divine foresight and divine providence in the arrangement of earthly life. He filled his multivolume works with reactionary, idealistic ideas that often blocked the road to a correct explanation of factual material. Therefore, Ritter's role as a systematizer of such material was negative rather than positive.

Although he adopted some features of Hegel's dialectical method, which caused the historical method to appear in his works, Ritter at the same time borrowed a number of Hegel's idealistic delusions that were subsequently discarded by Marx and Engels. For instance, Ritter regarded nature as a category whose development depended entirely on the level of development of reason or culture that had been attained at a certain time by mankind. He saw the chief difference of geography from other sciences in its study of the physical changes in our planet under the influence of human activity. This proposition would not sound bad if these changes were viewed as real changes that occurred under the effect of the forces of nature and the economic activity of human society, but Ritter viewed them as changes in the spiritual perception of nature by people as a result of their amelioration. He tried in his works to give concrete form to the well-known but highly unscientific attempt by Hegel to prove the reasonableness of Europeans' dominance over the inhabitants of other parts of the world. Hegel, as we know, affirmed that "the new world, in general, is an undeveloped dichotomy [America is meant— V. A.]: it is divided into a northern and southern part, like a magnet. The old world, on the other hand is a consummate dichotomy [sic] divided into three parts, of which one part, Africa, is a lode metal, a lunar element rigid from the intense heat, where man succumbs within himself; it is a mute spirit without consciousness. Another part, Asia, is a bacchanalian cometary frenzy, an environment violently born of itself, a formless work, without any hope of mastering its own environment. Finally, the third part, Europe, makes up the consciousness, the rational part of the earth, the equilibrium of its streams, valleys and mountains — and its center is Germany. The parts of the world are thus distributed not at random, not for the sake of convenience, but represent substantial differences."[29]

Guided by Hegel's quoted proposition, Ritter considered the character of the continents in direct connection with the life of their peoples and with state systems. In the process Europe and, above all, Germany, as predetermined by destiny, were credited with the most perfect form of life and primary significance in the life of all the other countries and peoples. "Thus, the extreme fragmentation and physical development of solid and liquid forms in the most confined space and the facility with which one can survey in Europe both natural relations and the everyday life of its peoples are what constitute the distinctive character of this part of the Old World. Just by its very nature, it seems to have a different designation from those parts of the world with which it has contact."[30] Moreover, Ritter, repeating Kant, contended that the geographical sciences should deal mainly with spatial categories, with the filling of space, with the description of filled spaces and their spatial relations. This is how Ritter distinguished geography from history, which limits its study to the investigation of things in terms of their successive development. Geography is the science of space and history, the science of development, Ritter said, repeating Kant's well-known thesis and propagandizing the view of geography as a spatial science thereby deprived of a material object of study. It is therefore not surprising that in his works he failed to apply with any success even the comparative method, which was well known to the scholars of his time. "Ritter restricted himself only to the comparison of external forms, he did not consider at all the genesis of various elements of the earth; he did not investigate transitional forms, which apparently link different elements, as does an anatomist or philologist or as modern comparative geography does on the basis of the investigation of different homologous forms."[31]

Having totally adopted Hegel's reactionary idea of historical and nonhistorical peoples, Ritter approached geography, on the one hand, as a purely spatial (chorological) science, and on the other, he carried on in his works a religious and natural-philosophic interpretation of the interrelations between nature and society. There took place here a kind of meeting of historical idealism and geographical determinism. The determinist geographers proceeded from a material basis, but, in attaching decisive importance in the life of society to the influence of the natural environment, they arrived at the inevitable conclusion that social development is predetermined, and that it could be predicted if the character of natural conditions were known. The idealists attributed the ties between nature and society to acts of God, who created nature as a basis for human life, which predetermines social development.

Thus, what would appear to be completely opposite schools in the development of geography led to an idealistic view of history. This phenomenon is highly characteristic of bourgeois science, in which inconsistent materialism often leads to idealistic conclusions. Such similarity in determining the significance of the natural environment's influence on the development of human society frequently gives grounds (to our mind, insufficient) for identifying Ritter as a representative of geographical determinism. This is wrong. Geographical determinism, although it did lead ultimately to a failure to understand the essence of the interrelations between human society and the rest of nature and to the mechanical transposition of the laws of nature to the sphere of social relations, still afforded an opportunity, as we have already pointed out, for a correct understanding of the natural environment as a single material whole. The idealistic theory, meanwhile, did not afford even this possibility. It was always an impediment to the development of science and led geography away from an understanding of the actual conditions of social development. It would therefore be a mistake to completely equate geographical determinism with idealism in all instances.

The positive contribution that the idealist geographers (including Ritter) made to science was not a consequence of their idealistic views but the result of using elements of the dialectical method, which enabled them to create integrated geographical representations and simultaneously to arrive at conclusions that were really very close to geographical determinism.

Ritter's works, despite their idealistic orientation, contain quite a number of profound and brilliant ideas. After removing Ritterian idealism (it is now at least humorous to rebuke the idealist Ritter for his idealism), one finds a great many correct dialectical propositions that are occasionally lacking in modern geographical works.

Ritter has a thesis about the interrelations between society and nature and on the profundity of these interrelations, and also on the objective character of the laws of nature, on the influence of human society on the development of nature, on changes in the natural appearance of countries and regions under the influence of man, on the influence of natural conditions on people's lives, and so forth. Although Ritter spoke of geography as a purely descriptive science, he urged geographers to rise to an understanding of the relations between spaces and to establish ties within the geographic phenomena occurring on the earth. " . . . The earth is independent of man; it has been the theater of natural phenomena before him and without him; consequently, the law of its forms and works cannot originate with him. One must search in the science of the earth itself for its laws,"[32] Ritter affirmed, despite his idealistic concept of the divine predestinationism of the process of development.

In touching on the ideological legacy left us by Ritter, I would like to cite a statement about him by D. N. Anuchin, which indicates the undoubtedly great influence that his ideas had on his contemporaries. "Although some of Ritter's views were essentially a revival of Strabo's views, they were based on a much greater amount of data of natural science and history and on a detailed critical analysis of geographic facts and gave a new meaning and life to geography as a science standing at the juncture of the natural and social sciences and seeking to link the course of world culture to geographic factors. Ritter is to be credited with arousing both in the scholarly world and in society, which were preoccupied at that time with metaphysics and natural philosophy, a new interest in geography as a science that was not only physical and mathematical but also philosophic and historical."[33]

In our opinion, although Ritter was one of the spokesmen of idealism in geography, he must nevertheless be included among those scholars whose work merits special, meticulous study and scientific analysis.

Ritter had a noticeable influence on a segment of Russian geographers. Thus, the instinctive materialism that serves as the theoretical basis of the anthropogeographical works of A. A. Kruber, L. D. Sinitsky and V. P. Semenov-Tyan-Shansky was combined by them with elements of idealism. The same may be said of works by members of the French Géographie Humaine school, in which exaggeration of the importance of the natural environment in the development of human society is accompanied by a pretty idealization of the terrestrial whole as some harmonious unity without contradictions. In retouching geographical determinism with dialectical idealism, they arrive at the conclusion that the natural environment influences society, but social life itself is depicted as a harmonious, classless unity. Still, it should be noted that neither the members of the Géographie Humaine school nor those of the Russian anthropogeographical school reached the reactionary conclusions that the German anthropogeographers — Ratzel, in particular — did. The work of the French anthropogeographers (like that in the works of A. A. Kruber, L. D. Sinitsky, and V. P. Semenov-Tyan-Shansky) contains many correct materialistic propositions and interesting facts that show the relationships within the geographic environment between society and the rest of nature. However the discussion of these relationships is marred by its one-sidedness: attention is focused on man's passive adaptation to nature, on nature's influence on man, yet almost nothing is said about his overcoming unfavorable natural conditions. To our mind, both the French and the Russian anthropogeographers took a step backward by comparison with Reclus and Mechnikov.

This was correctly noted by Yu. G. Saushkin, who writes: "Reclus and his group regarded man as a *toiler* who changes the face of the earth in the process of labor, whereas Vidal de la Blache and his followers, as well as German, Russian, and other anthropogeographers, began to regard man only as the *inhabitant* of the earth."[34] Indeed, whereas in the works of Reclus and Mechnikov we encounter certain errors based on geographical determinism, in the works of the anthropogeographers materialism has already degenerated into geographical determinism. The works of the anthropogeographers mark the end of the period in which writings based on geographical determinism were still scientific in character. The subsequent development of the concepts of geographical determinism resulted primarily in manifestly unscientific conclusions.

Beginning with Humboldt and Ritter, the disparity between

the two trends in the development of geography intensified noticeably. One segment of geographers went into natural science, in which many valuable investigations of individual components of the natural environment were conducted, primarily on the basis of instinctive materialism. This segment of geographers began increasingly to shun questions related to population geography and economic geography. In those instances in which these questions were touched upon, these geographers, with rare exceptions, failed to rise above geographical determinism.

This departure into the study of the individual components of the natural environment is easy to explain. The naturalist geographers of the late nineteenth century (and those of the early twentieth even more) began to understand the untenability of geographical determinism most clearly in their attempts to comprehend the geographic environment as a whole.[35] However, most of them were unable to accept the idealistic concept of the unity of the world, since it contradicted their basically materialistic world outlook. Marxist philosophy, which affords the possibility of a scientific understanding of the interactions between nature and society, meanwhile, had not yet been added to the general arsenal. It was for this reason that geographers now wished to depart from the study of social elements of the geographic environment and from consideration of insoluble questions. There was an extreme intensification of the process of isolation of physical geography, which started to be called a purely natural science not at all related to the social branches of geography. Numerous assertions appeared to the effect that physical and economic geography were completely different sciences, and some economists even began to exclude economic geography from the system of geographical disciplines.[36]

Despite its outstanding achievements in the understanding of the natural environment and especially of its individual components, the further development of this school inevitably leads to the affirmation of a cleavage in the unity of the material world, and therefore turns into a hindrance to further comprehension not only of the landscape envelope as a whole but also of its individual components. Comprehension of the individual elements of the natural environment becomes impossible when the influences of human society on them is completely ignored.

In our opinion, not even physical geography in its future development can be based on natural science alone. But it is especially important that such a departure from geographical objects of study that are social in character and the placing of a wall between physical and social geography leads to the total impossibility of comprehending the earth's landscape envelope as a whole. This is also the reason, incidentally, that certain materialist geographers of the past, even while seeing the untenability of geographical determinism, continued to adhere to it. Indeed, a repudiation of geographical determinism, with a concurrent denial of the universal interconnection between phenomena and with the conclusion that the development of geography as a particular science is impossible, represents a long step backward toward indeterminism.

It is very typical that none of the most prominent Russian naturalist-geographers of the past took this step backward. For the most part they never affirmed the existence of a gulf between physical and social geography, although it actually appeared in applied research, which was carried out almost exclusively in the area of the natural complex of elements of the earth's landscape envelope. The research by this group of scientists was a major contribution to geography, but almost exclusively to physical geography. The social elements of the geographic environment remained almost entirely outside the realm of study of the materialist geographers, who, as a rule, strove not to go beyond the bounds of natural science. Although certain scientists came close to a correct understanding of the unity of the earth's landscape envelope, they were unable, from positions of an instinctive materialism that was not entirely without dialectical elements, to find the right path to an understanding of its social elements and were unable to determine completely their qualitative features and specific laws of development. V. V. Dokuchayev (1846–1903), A. I. Voyeikov (1842–1916), D. N. Anuchin (1843–1923) and, abroad, F. Richthofen (1833–1905) were the most prominent representatives of this classical pre-Marxian geography, and their works retain most of their value to this day.

Another group of geographers, who also accepted dialectics in some measure, tried to apply it on an idealistic basis. In the concrete study of geographic phenomena, such dialectics could yield very little of value. It is therefore not surprising that the members of the idealistic school contributed substantially less to physical geography, and what they did contribute was not due to, but in spite of, the idealistic basis of their world outlook. They did even less in the study of specific geographic phenomena of a social character, usually failing to rise above anthropogeographical concepts, and in this respect, incidentally, they resembled a large segment of materialist geographers.

The positive aspect of the use of dialectics, even in its idealistic form, was that a transition from indeterminism to historical idealism became feasible. Dialectics helped many idealist geographers (Ritter, for example) to abandon the disjointed examination of nature and society that had always been a concomitant of geographical indeterminism. In visualizing the world as a developing, definite whole, some of the geographers who adhered to positions of historical idealism even attained an understanding of the qualitative difference between nature and society within this whole.

But since the dialectical concepts of that group of geographers were simultaneously idealistic, they greatly distorted concrete geographical material. The idealist geographers attributed the unity of the world, as well as the differences within this unity, not to the unity of matter but to the existence of a developing, single spiritual foundation that was subordinated ultimately to the will of divine providence, which is unintelligible to man. The integrated character of the idealistic theories was merely a subjective fact that did not reflect any external reality. They regarded nature as something bound up through fate with the history of the peoples populating the earth, and the specific character of social relations continued to be linked to the direct determining influence of natural conditions.

That is why dialectical idealism (or, in other words, idealistic monism), although something of a step forward in geography, could not bring geography to an actual understanding of its object of study. This school can be called progressive only in comparison with indeterminism, but it was unable, of course, to equip geography with an authentically scientific method of comprehending the material world.

Incidentally, it is difficult to find among nineteenth-century geographers even one major scientist who was a completely pure idealist and adhered totally to positions of indeterminism and idealism. The work itself with concrete geographical material led idealist geographers, despite their theoretical views and concepts, to certain materialistic inferences and conclusions. One can find materialistic propositions in the work of almost every geographer who believed in a supreme intellect or spiritual cause.

"Natural science is materialistic in substance; materialism and its roots lie in nature. Natural science instinctively gravitates toward dialectics. To avoid erroneous concepts in learning, one must know the only correct philosophy, the philosophy of dialectical materialism . . . .

". . . The objective world — nature — is preeminent; man is a part of nature, but he must not only contemplate this nature externally, he can, as Karl Marx said, change it."[37]

This instinctive gravitation of natural science toward materialism and dialectics undoubtedly played a large role in the development of geography, also explaining why the idealist geographers arrived at materialistic, and sometimes dialectical, conclusions.

The materialist geographers of the nineteenth century, in turn, made a great many separate statements of an idealistic kind, especially in those instances in which they attempted to ascertain the essence of geographic phenomena of a social character or the interrelations between society and nature. Dualism, thus, was quite widespread in the theoretical concepts of many geographers until they began to switch to positions of dialectical materialism. Dualism was not the least conducive to the development of geographical theory and hampered this development.

On the whole, however, the influence of materialistic philosophies has had appreciably more impact than the influence of idealism. This influence may be observed in literally every school of pre-Marxian geography. Materialism had a strong influence, we repeat, even on geographers who basically held an idealistic view of the world. The difference between the basically materialistic geographical determinism and the idealistic school stems from the differences in outlook and reflects, on the one hand, the inconsistent materialism of bourgeois philosophy and, on the other, varieties of idealism.

To fail to see this fundamental difference is to fail to understand the charcter of the individual stages of development of geography as a science. To affirm that geographical determinism and idealism are equally reactionary in the history of geography on the grounds that geographical determinism has now degenerated into a variety of idealism is to commit nihilistic errors that make it impossible to absorb and utilize correctly the theoretical heritage left to us by past geographers. It is perfectly clear that if there is to be a successful scientific analysis of the works by geographers of the past, it is important first of all to ascertain the *basis* (materialistic or idealistic) or these works and the philosophic basis of their theoretical concepts. Everything that geographers of the past did in the realm of theory is sometimes declared unscientific and pernicious in advance, which, incidentally, "eliminates" the need to study their theoretical concepts.[38] In our opinion, this type of attitude toward the geographers of the past is a most harmful phenomenon.

Knowledge of the general laws of development of nature, society, and human thought, which are specially studied by philosophy, equips scholars in all fields with a correct world view, something that is of inestimable value for orientation in practical activity and in all areas of the scientific investigation of matter. However, knowledge of general laws does not furnish ready-made solutions and does not exempt representatives of the concrete sciences from a theoretical study of the objects and phenomena being investigated. Attempts to deduce physical, biological, geographical, economic, and other laws *directly* from the general laws of dialectical materialism can lead to the most glaring errors. It should be remembered that the laws of the concrete sciences *are not* merely simple, particular manifestations of already known general laws of dialectical materialism. General philosophic laws do not exist independently of objective reality, just as the general does not exist without the particular. Knowledge of general philosophic laws does not make knowledge of particular, concrete patterns superfluous. Moreover, knowledge of physical, biological, geographical, economic, and other laws is a highly important condition for the development of philosophy, which depends on the development of the concerete sciences to the same degree that the concrete sciences depend on the development of philosophy.

The theory of Soviet geography is based on Marxist-Leninist philosophy, and the works of the founders of Marxism-Leninism are of exceptional and decisive importance to us; but this does not give us the right to ignore the theoretical opinions of bourgeois scholars. In our view, Soviet geography would unquestionably benefit from specialized scientific research that critically analyzes the works both of the most prominent geographers of the past and of contemporary foreign geographers who attempt to write theoretical works without relying on Marxist-Leninist philosophy. Such research would be of great assistance not only toward a correct understanding of the history of geography, specifically the history of the development of geographical ideas, but also to the development of its theories.

In summarizing our examination of the two trends in the development of geography, it should be pointed out that ultimately neither one offers any further prospects of development for geography. The first, basically materialistic trend leads in the course of its development to a loss of understanding of the unity of geography's object of study, an understanding that the first representatives of that trend still had. Although providing temporary, relative development of certain subfields of physical geography, it impedes, in the long run, the use of synthesis and leads to the unintelligibility of the landscape envelope of the earth as a whole.

The second trend, the idealistic one, combines all the elements of the earth's landscape envelope, considering their development to be subject to the action of natural laws alone, which are controlled by a supreme being. The mechanical carry-over of the laws of nature into the sphere of social relations results in the impossibility of comprehending not only the whole but also the parts, that is, the individual elements of the landscape envelope. The idealist geographers often even declared the world unintelligible. And recognition of the unity of nature and society in idealistic form (as a reflection of a single divine spirit) is not capable of ensuring the study of this unity, since this recognition loses its materialistic essence.

Consequently, both schools, the instinctively materialistic one and the idealistic one, join together in their development, lead to the liquidation of geography as a science, and come into direct conflict with the laws of materialistic dialectics.

The history of geography has seen examples in which certain geographers, on the basis of old, idealistic, philosophic concepts, endeavored to create a theory of unified geography. Most often these efforts were made in terms of denying the qualitative differences between the laws of nature and the laws of social development.

These efforts may be said to include the works of Ratzel (1844–1904) and the theoretical constructs of Hettner (1859–1941), as set forth in his work, *Geography: Its History, Character and Methods.*

Largely repeating Ritter's reasoning and regarding himself as Ritter's pupil, Ratzel tried to create the concept of a unified geography by proceeding from recognition of the causation of social development by geographic factors. There is essentially nothing new in his theoretical discussions. In fact, he pointed out himself in the work, *The Earth and Life,* that he was only developing Ritter's ideas on the character of natural regions in their relation to the life of peoples and in relation to political geography. But, in repeating Ritter's ideas, Ratzel developed their most reactionary propositions. Ratzel sees the geography of society primarily as a division of biogeography. He even eliminates the distinction between society and nature that Ritter always acknowledged, albeit in idealistic terms.

"Man's relation to the soil is the same as the relation of all living things to it. The universal laws of the proliferation of life include as well the laws of the proliferation of human life. Therefore, anthropogeography is conceivable only as a branch of *biogeography,* and a whole host of biogeographical questions may be carried over directly to questions of the proliferation of man."[39] According to Ratzel, social phenomena are not only attributable to the influence of natural conditions, the study of the development of nature itself must be approached from the standpoint of political tasks; in other words, any geographical inquiry was totally subordinated to political tasks. The objective character of the geographical study of the landscape envelope was denied.

Ratzel's concepts contained all of the basic theses that were developed later by representatives of the pseudoscience of geopolitics. Physical geography was viewed as a science that was subordinated completely to politics, and the natural environment was treated as a natural justification of political aggression. Ratzel likened every state to a living organism with an inborn urge to capture territory. This latter proposition was declared to be a law-governed natural phenomenon characteristic of all living things; wars, consequently, were not the result of imperfect social organization but the manifestation of an inherent property of man. "In the history of mankind, the urge to capture as much territory as possible has been one of the most powerful forces, and when we see throughout the history of past centuries and in the life of contemporary peoples that this urge to capture and retain as much territory as possible continues to develop, it is merely a recurrence of that which has transpired thousands of times already in the vegetable and animal kingdoms."[40] It was statements of this type that were widely picked up by the most reactionary bourgeois sociologists, for whom the combination of idealism with naturalism has perhaps become one of the most characteristic traits. They regard the whole process of social development merely as adaptation to the natural environment, which allegedly is the only thing that is capable of explaining and substantiating everything.

By making use of Ratzel's ideas on the geographic causation of political life, Rudolf Kjellen and Karl Haushofer (president of the Academy of Sciences of Germany during the period of fascism) created the pseudoscientific theory that was given the name of geopolitics (the term *geopolitics* was first put into use by Kjellen).[41]

The British geographer Mackinder (1861–1947) also eschewed the interpretation of geography as an autonomous science in his theoretical concepts, subordinating it completely to politics. The unity of geography, according to Mackinder, is based above all on the subordination of all geographical problems — physical, biological, and social — to the influence of politics. Without rejecting inquiry into the essence of phenomena, he sees his purpose primarily in substantiating the facts of political life. Geography, according to Mackinder, should deal with man's interrelations with his natural environment, with the understanding that they constitute a single living organism. He did not consider it his main purpose to show the peculiar characteristics in the geographic conditions of individual countries and regions and did not seek to develop regional geography, but wanted to show human history as part of the life of the world organism.

In doing so, he viewed human society as a combination of alliances that formed in the process of the struggle for existence, and nature as the basis of political phenomena. For example, soil and climatic conditions supposedly determine population density, and a comparatively high population density was regarded as the chief condition for the development of civilization.

Mackinder considered "geographic inertia," that is, geographic causality, to be the decisive factor of social development. The geographical situation, the importance of which, according to Mackinder, increases as a country develops, was declared to be the determining and most basic thesis of geographic inertia. He saw the population only as a part of nature, as a will-less mass completely subordinated to the effect of the law of geographic inertia. In attaching such exaggerated importance to geographical situation, Mackinder contended that possession of the eastern part of Europe assured authority over the "pivotal region of the earth" — the center of Eurasia — and, consequently, it afforded the possibility of establishing world domination.

Mackinder's concepts were widely used by the geopoliticians, and they had an influence on some foreign geographers who were not directly involved with geopolitics. For example, certain present-day American physical geographers consider it necessary to subordinate their research entirely to political goals. Viewing geography from Mackinder's position, they hold that "physical geography furnishes [politicians — V. A.] the thoroughgoing scientific conception of the earth that is essential to rational strategic and tactical planning,"[42] and for this reason it is a political discipline.

By preaching the unity of nature and man, by proceeding from a denial of the qualitative peculiarities of human society and by transforming physical geography into an applied political discipline, the bourgeois geographers and sociologists of this school gradually do away with any semblance of an objective study of the earth's landscape envelope. It is precisely scholars of this type that are meant in Engel's phrase about people in science who ". . . now find that at least in this area [political economy — V. A.] it is safest not to *tolerate* any science at all."[43] This area of "intolerance of science" now includes many scientific-like theoretical views that are widespread in modern bourgeois geography. Indeed, can explanations of the essence of social phenomena that are constructed on a complete denial of human society's qualitative difference from the animal world really be considered scientific? "It must be remembered that physically man is in essence an animal, subject to the same laws as the rest of nature, and that the multitude of traps that other societies have fallen into in the process of their evolution are also lying in wait for human society."[44] What is meant by "other societies" here is ants, bees, and so forth; in fact it turns out that ants have a division of labor, exchange, slavery, a colonial policy, and many other social phenomena observable in human society.

Modern bourgeois geography still uses various combinations of old, long-ago bankrupt theoretical concepts, which are presented as something new. Most often all of these new theories evince a mixture of historical and even subjective idealism with naturalism and geographical determinism.

An especially large number of such works exist in American geography. Thus, Huntington, in summing up his theoretical research,[45] defined as the chief driving forces of civilization: (a) biological heredity; (b) cultural talent; and (c) geographic environment. In short, a full opportunity is provided here for constructing the most variegated racist, Malthusian, and geo-political pseudotheories. The authors of the well-known collection, *Geography in the Twentieth Century,* edited by E. Taylor, attempt to reject completely the disclosure of any ties between nature and society.[46] Even the ties between the local characteristics of production and the conditions of the natural environment are ignored. Perhaps most typical in this respect is the article by R. Platt, "Determinism in Geography,"[47] which, based on a correct acknowledgement of the untenability of geographical determinism, proposes that the influence of nature on human society be disregarded altogether. Platt holds that the factors determining the life of society lie in the special properties of human blood, racial characteristics, and so forth, and that no evidence need be cited for his theoretical propositions, since they should, it turns out, be taken on faith. According to Platt, there is an obvious advantage to operating with plausible assumptions and setting forth truths that do not require proof, after which one may look for the most acceptable answers to given questions. In essence, geographical determinism is criticized from the indeterminist standpoint of American pragmatism, that is, a step is made not forward but backward with respect to geographical determinism, which turns out to be too progressive for a certain segment of present-day representatives of bourgeois science.

Hettner, who upheld a unified geography, based his theoretical concepts on geographical determinism. He adopted Humboldt's materialistic ideas and used them in the study of landscapes, viewing the latter as real, objectively existing phenomena.

But, in addition, his world view was significantly influenced by the idealistic philosophy of Kant and especially Hegel.

These influences of idealistic philosophy were most evident when Hettner attempted to develop general theoretical propositions about geography as a whole.

In trying to determine the place of geography in the general system of sciences, Hettner could not find the right way to solve the problem. Although he upheld the unity of geography, he failed to see the unity of the material object of study of geography. Hence his rejection of the subject principle of the classification of sciences and his theory proposing the classification of sciences without reference to the material objects they deal with. The novelty of this theory was illusory. In effect, Hettner tried to reestablish Kant's classification of sciences: he declared geography to be a particular spatial (or chorological) science dealing with spatial relations. "But geography should not be a science of the distribution of various objects among localities, but a science of the filling of space. It is a *spatial* science in the sense that history is a temporal science."[48] In upholding geographical determinism through his acceptance of the idealistic Kantian concept of divorcing space from time, Hettner vehemently objected to the geopolitical interpretations of geographical determinism. Hettner's world outlook was bourgeois. His theoretical conception of geography did not withstand the test of time. But calling him a "mouthpiece of the misanthropic ideas of imperialism" is also a great exaggeration that is very misleading.[49]

Hettner came out against Ratzel, commenting disapprovingly on his "natural regions of physical geography." He regarded as wrong the effort to turn physical geography into a utilitarian discipline of politics, into a science only of "man's habitation." "The nature of countries exists above all for itself and should be studied and accepted for itself. Man develops in nature and in dependence on nature."[50]

In establishing the chorological concept of geography, Hettner rejected the principle of a comprehensive study of the earth already advanced by Varenius. Although Hettner raised the comparative method to an absolute, he failed to comprehend the qualitative characteristics of the process of social development; thus he concluded that it was impossible to understand the geographic envelope of the earth as a whole, since there was nothing with which it could be compared. Therefore, Hettner suggested excluding earth science from geography. The object of study of geography, in his opinion, should be limited to the study only of local characteristics; in other words, geography was reduced only to its regional aspect, which he called regional geography. In the process it was overlooked that regional geography, as a science dealing with the parts of a whole, could not develop without its other division, which dealt with the same object of study as a whole. It need hardly be proven that without a study of the general laws of development of the earth's landscape envelope, the individual landscapes — parts of this envelope — cannot be understood. This absolutely correct proposition had been advanced in general form previously by Varenius and was not disputed by anyone before Hettner.

Hettner reduced geography to the study only of individual complexes and landscapes that formed on the earth's surface. However he viewed the landscapes as complexes within which the social elements were conditioned completely by the laws of nature. This resulted in lumping together elements that were developing under the effect of fundamentally different laws and eventually in the impossibility of understanding not only the earth's landscape envelope as a whole but also individual landscapes.

Hettner's theoretical concepts, in general, contain many contradictions. By declaring space a special object of study and divorcing it from time, he strays from the materialistic view of the world in the direction of idealistic monism. At the same time Hettner affirms quite rightly that "only through history can we understand the present." Furthermore, Hettner was also basically right when he opposed the attempts of some physical geographers to completely isolate the geography of nature from the geography of human society. For example, he refuted the views of those American geographers who held that geography should be limited to the study of the earth's solid envelope. "This view, which is very possibly rooted in American research works on the Cordilleras, is still alive and, under the influence of an American, Davis, has been becoming more widespread in Germany as well. However from its very first steps it has come into conflict with the historical development of science, in which the study of the earth's solid envelope has always played an important, but never a decisive, role."[51] With even more vigor Hettner opposed the geographical indeterminism that was being preached by members of the Baden school of Neo-Kantians. "Geography cannot be limited to any definite range of phenomena of nature or human life; it should embrace at once all the kingdoms of nature and man as well. It will be neither a natural nor a social science — I take both these words in their usual sense — but the one and the other combined,"[52] Hettner wrote.

Hettner's overall system of views, which is not without dialectical elements, cannot therefore be considered merely a reflection in geography of Kant's philosophy. In the domain of concrete research Hettner always adhered to materialistic positions. In short, the geographical research Hettner did placed him in conflict with his own concepts. Hettner, the empiricist, rebelled against Hettner, the theorist. In working out concrete questions of geography and, in particular, questions of regionalization, Hettner approached them realistically, diverted from his philosophic views. Contravening his general theoretical concept, he correctly affirmed that "the task of geography, like that of any other science, is to understand reality as it is."[53]

In taking note of the valid propositions in Hettner's theories, which are of indisputable positive value in the fight against indeterminism, one must say at the same time that they are unable to compensate for the scientific untenability of Hettner's theoretical concepts as a whole. The objects of study of any concrete science are various concrete forms of matter that are in constant motion (development) in space and time. Matter does not exist outside of space or time and can be understood in its concrete forms only under the indispensable condition of *simultaneous* investigation in time and in space. Any concrete science deals not only with the origin and development of its object of study but also with its spatial situation (which is why there cannot be a science of location). A science with the purpose of studying *only* spatial forms and relations can exist solely in the realm of abstract thought, in the form of "conscious abstraction" from all concrete properties and qualities; examples of this kind of abstract science are geometry in mathematics and kinetics in mechanics.

But geography is a concrete science. It has a totally concrete, material object of inquiry. Thus, to speak of geography as an abstract, purely spatial science is to make a serious mistake.

In summarizing our brief acquaintance with the history of the basic ideas in the geography of the past, it must be pointed out that theory in geography was connected in one way or another with philosophy throughout time and in every country. The theoretical concepts of geographers always mirrored the views of certain philosophers, while the philosophers, in turn, made wide use of geographical materials to substantiate their views.

The struggle within philosophy between the materialist and idealist schools manifested itself in geography chiefly in the form of the struggle of determinism, which reflected the views of materialist philosophy, against indeterminism, which reflected the views of idealist philosophy. The philosophy that was progressive for its time was always the basis for the theoretical concepts of the progressive geographers who moved their science forward. Conversely, idealistic and reactionary philosophic concepts were invariably used in geography as the bases for various unscientific distortions. This proposition remains valid today. For example, the revival of various trends and schools of idealist philosophy that may be observed now in many capitalist countries has become the basis for the revitalization and intensification of indeterminism in bourgeois geography.

The erroneous aspects of the materialist philosophers' works that were generally progressive for their time invariably led to incorrect definitions and conclusions in questions of geographical theory. Thus, the inconsistency and mechanistic character of the materialistic bourgeois philosophy caused the emergence of geographical determinism, which in its development then joined with reactionary indeterminist views. In this way, inconsistent materialism (in the form of mechanistic determinism) leads to idealism not only in philosophy but also geography.

CHAPTER FOUR

The Reinforcement of Empiricism with
Inadequate Synthesis. The Significance
of the Works of V. V. Dokuchayev and
D. N. Anuchin for Geographic Synthesis.
Empiricism in Geography and the
Training of Geographers in Universities.

In the second half of the nineteenth century, the development
of geography entered a period not only of an intensified debate
in the field of theory but also of continued and even sharper
differentiation of geography. This was occasioned, first, by
the opportunity of turning from basically descriptive forms to
a more intensive study of the nature of individual phenomena
and, second, by the absence of a theory that afforded the
possibility of a scientific understanding of the landscape envelope as a whole. The old idealistic and inconsistently materialistic concepts were demonstrating more and more their
scientific untenability. The necessary conditions did not yet
exist for the creation of a new theory based on the principles
of materialist dialectics.

It was no accident that V. P. Semenov-Tyan-Shansky
wrote that "... the 1860s saw the beginning of the intensified
dismemberment of geography into separate, unrelated disciplines, and geography began uncontrollably to consume itself
in the disclosure of nothing but bare facts .... The concept of
the geographic landscape as a law-governed spatial complex
of objects and phenomena on the surface of the earth has been
dispelled."[1] Indeed, geography started literally to creep in
different directions and to turn to the study of an infinity of
totally unrelated details.

The new branches that were constantly emerging were beginning to develop largely on their own. In geographical research on individual subjects (relief, climate, agriculture, in-

dustry, etc.,) that did not take proper account of the interaction between them, geographers sometimes ceased to understand one another. As a result, the lag in the creation of integrated geographical accounts of countries and regions increased. There were fewer and fewer geographers capable of composing such accounts, even in purely descriptive form; meanwhile, the need for works of this kind never disappeared.

Thus, it may be considered that in the middle and especially at the end of the nineteenth century there was a transition from the synthesis that had been carried out in the first half of the nineteenth century in various forms by several geographers (Humboldt and Ritter, for example) to more intensive analysis. Unable to form a correct understanding of the unity of the material world, geographers were forced to confine themselves to intensive study only of individual elements of the earth's landscape envelope. This analytic trend in geographical research extended its framework in the direction of more intensive study of details, which in turn necessitated a more clear-cut classification of the objects and phenomena geography was concerned with, in accordance with the character of their properties. This required a specialized approach to each subject, that is, the creation of special methods for its study, and this was of unquestionable value and became one of the forms of development of geography.

The strengthening of analysis without proper synthesis during that period led to the declaration: "The time of the Humboldts has passed." Thus the development of geography was limited to its subfields or, as we often call them, the individual geographical sciences.

This path of development has both positive and negative aspects. The twofold significance of differentiation was pointed out by Engels, who discerned it in the development of science from the sixteenth through the nineteenth centuries: "The decomposition of nature into its individual parts, the division of the various processes of nature and natural things into certain classes and the investigation of the internal structure of organic bodies — according to their diverse anatomical forms — were all major conditions for the gigantic success that has marked the development of natural science over the past four centuries. But the same method of study has left us with the habit of considering the things and processes of nature in isolation, without their great common link, and, because of this, [we consider them] not in motion but in an immobile state, not as something changing substantially but as something eternally immutable, not as something alive but as something dead.''[2]

To a considerable extent the preponderant development of individual geographical sciences was also promoted by the needs of life, since these sciences are capable primarily of satisfying *direct* economic needs, whereas general geographical work is mostly of *indirect* value for applied work, and therefore its necessity is not always realized.

Yet it would not be quite correct to say that geography began in the mid-nineteenth century to develop exclusively in the direction of analysis. This tendency was distinctly dominant and continues to be dominant today. But the necessity of synthesis, the necessity of creating integrated pictures of the landscape envelope that surrounds us, was clear to certain scientists in that period as well. Moreover, it may be asserted that there has not been a period in the past five hundred years in which there was no understanding at all of the necessity of synthesis.

In studying certain aspects of nature, naturalists sometimes arrived at an understanding of the need to investigate their ties with other components of nature and to investigate them as parts of a more complex whole. When relief, climate, soils, and so forth begin to be studied as parts of a whole, as elements of the earth's landscape envelope, the geologist inevitably will come to geomorphology, the meteorologist to climatology, and the soil scientist to physical geography. Thus, whereas some geographers, in crossing the boundaries of geographical analysis, as we shall go on to show, lost their understanding of the common ties between the objects of study of the geographical sciences and at the same time departed from geography, certain representatives of sciences associated with geography, in turning from the analysis of objects of study to synthesis, went in the opposite direction and came to geography, although often they themselves did not fully understand this fact and, in an organized manner, remained outside of specialized geographical institutions. Thus there appeared a rather curious phenomenon, still current today, in which specialists calling themselves geographers are essentially *not occupied with geography,* while certain representatives of sciences associated with it who do not consider themselves geographers are enriching *geography.*

A concrete and more vivid example of the geographical work of a naturalist who did not consider himself a geographer is the work of the great Russian scientist V. V. Dokuchayev.[3] In studying soil as a perpetually changing function of a number of other components of nature, Dokuchayev came close to a correct understanding and definition of the substance of geography when he spoke of the field that was taking shape. "Situated by *its very nature,* one may say, in the very center of *all of the most important* divisions of natural science, which are geology, orohydrography, climatology, botany, zoology, and, finally, the study of man in the broadest sense of the word, and thus naturally drawing them *closer together* and even *connecting* them, this discipline, which is still very young but is filled with the highest scientific interest and significance, is achieving more and more successes and gains with each passing year . . . and the time is not far away when it will *rightfully* occupy, by virtue of its *great importance for the destiny of mankind,* a fully autonomous and distinguished place, with its own rigorously defined *tasks and methods,* without becoming merged with the existing divisions of natural science or, even less, with the geography that is branching off in all directions."[4]

The last remark is of specific interest to us geographers. Indeed, a geography "torn asunder," which is a variegated conglomerate of essentially disjointed sciences without a common theory, without a common object of study and a common method, is not at all consistent, of course, with the field outlined by V. V. Dokuchayev. But there is no substantial difference whatsoever between V. V. Dokuchayev's field and geography in its *scientific* interpretation.

Dokuchayev viewed his field within natural science, as though he regarded it as one of the natural sciences. This indeed is the customary view. But his inclusion of geography in natural science is actually based on a broader interpretation of this area of human knowledge than is common today. If human society is excluded from natural science, then Dokuchayev's inclusion of geography in natural science becomes purely nominal. The reason for this is that in his theory Dokuchayev also included man "in the broadest sense of the word." The unity of nature, therefore, also presupposed societal elements, and one can hardly regard as fortuitous Dokuchayev's distinction between the new field and all the other divisions of natural science. Obviously one can say that *by its very nature* the study of the laws that govern the numerous and heterogeneous interrelations and interactions between nonliving and living nature, on the one hand, and human society on the other, cannot possibly fit into the framework of dehumanized natural science.

Yet Dokuchayev's theory is usually interpreted in geographical literature merely as a scientific substantiation of physical geography. Dokuchayev's mention of "man in the broadest sense of the work" is regarded as a token of esteem to geographical determinism. This interpretation, although it has formal grounds, cannot be considered valid. Of course, if one places physical and economic geography under different scientific systems, if one denies the dialectical unity in the object of study and the methodological basis that really is common to all of the geographical sciences, then Dokuchayev's allusion to man will indeed have to be considered erroneous. But if one views geography as a science of the earth's landscape envelope, which consists of a multiplicity of correlations and interrelations, including the interaction between society and nature, then one cannot help but recognize that Dokuchayev came extremely close to a correct scientific definition of the essence not only of physical geography but also of geography as a whole. We contend, therefore, that

Dokuchayev, by scientifically substantiating a system of geographic natural zones and creating a modern physical landscape science, developed later in the works of L. S. Berg and his pupils, made a major contribution not only to physical geography but also to the theory of geography as a whole.[5]

Dokuchayev's theory of the interaction between all the elements of living and nonliving nature and between nature and society is a highly important dialectical proposition that must not be overlooked in efforts to clarify the object of study, method, and basic tasks of geography.

Dokuchayev's theory appears to us to have been one of the first attempts at a scientific, theoretical substantiation of modern geography, which, by the very nature of its object of study, is forced to go far beyond the bounds of natural science and include social sciences as well. Thus Dokuchayev's inclusion of man in the material world of nature indicates that he correctly understood the essence of the relations between society and nature and did not try to sever man, as a special, spiritual substance, from nature.

"The chief objects of study have been *individual* bodies — minerals, rocks, plants, and animals — and phenomena, *individual* elements — fire (volcanism), *water, earth,* and *air,* in which, we repeat science has achieved astonishing results; but not their *relationships,* not the *genetic, everlasting,* and always *law-governed* connection that exists between *forces, bodies,* and *phenomena,* between *nonliving* and *living* nature, between the vegetable, animal, and mineral kingdoms, on the one hand, and man, his everyday life, and even the spiritual world, on the other. Yet it is precisely these *relationships,* these law-governed *interactions,* that constitute the *essence of the understanding of nature,* the nucleus of true natural philosophy — the best and supreme beauty of natural science. It is these, as will become clear below, that should form the basis for the entire structure of human life, including even the moral and religious world ...."[6]

This passage shows that Dokuchayev distinguished human society as a special category within the material world of nature and proposed examining the ties, on the one hand, between the components of nature and, on the other, between their organic combination and man. Without severing man from nature, Dokuchayev still contraposed him to the rest of nature, within nature, as a particular sphere requiring specialized *theory,* that is, the social sciences, for its understanding. The quoted passage also shows that Dokuchayev saw the geographic environment (and his "relationships and interaction" can only be interpreted as meaning the geographic environment) as the basis of "the entire structure of human life." Of course, if one views the geographic environment as the basis of change, as the basis of the development of human society, then this kind of assertion is wrong. But the point is that Dokuchayev, in considering the importance of the geographic environment, speaks of it as the *principal condition for the development of society.* In fact, that was the title of one of his works, in which he develops certain basic propositions of his theory: *The Primordial and Everlasting Conditions of the Life of Man and His Culture.*[7] If by geographic environment one understands the *conditions* of development, then they do in fact form "the basis of the structure of human life." Dokuchayev assumed the environment to be a necessary external condition of development, which is absolutely correct.

Nowhere did Dokuchayev ever say that his theory is a theory of the *cause* of social development, and he *did not carry over* the laws of nature into the sphere of social relations. Moreover, in examining the process of man's struggle against nature, Dokuchayev referred to the signficance of the *social system* in this struggle against nature, although he did not elaborate specifically the question of the essence of social relations or of the essence of interrelations between nature and society.

Here is a passage from Dokuchayev that is additional evidence of his understanding of the *qualitative* characteristics of social categories, the fundamental differences between human society and the rest of nature, and the significance of social structure in the process of man's struggle against nature. "Is it possible to prove *historically,* quite *precisely,* that the number of *slaves* of nature and of *the social system* has decreased in the past one hundred fifty years by so much as half a percent? On the contrary, has this horrendous number not increased on account of the new, contemporary and perhaps the most evil and ruthless phenomenon — capitalism, which is economic and industrial servitude . . .?"[8] Here we can ask the question: After reading these lines, can one consider their author a spokesman for geographical determinism? We think not.

Dokuchayev's efforts to substantiate geography as a science not only of the components of pure nature but also of the interactions between nature and society, including social elements in its objects of study, may be regarded as evidence of his perspicacity. Without having at his command the tremendous scientific discoveries made by Marx and Engels, he was able, from a position of instinctive materialism, to perceive quite correctly, albeit in general and vague terms, one of the principal manifestations of the general laws of dialectical materialism, to approach a correct understanding of the nature of the earth's landscape envelope, and to propose it as the object of specialized scientific inquiry. This, in our opinion, is the great contribution of V. V. Dokuchayev as a theorist of geography as a whole.

The basically correct definition of the subject matter of geography furnished by V. V. Dokuchayev is one of the important theoretical bases of the development of geography — a science that is mainly synthetic by the very nature of its object of study and that deals with the environment that has evolved on the earth's surface under the influence of interactions between nonliving and living nature, and between nature and human society as a special qualitative category of that nature. In studying its common object, geography views nature not only as the result of interactions but also as a system carrying out a process of interactions, as an environment in which production develops and which, while changing in the process of this production, itself influences changes in the life of human society by means of the same process of production.

Dokuchayev's role in the development of theoretical concepts in geography and in the individual geographical sciences is difficult to exaggerate. We believe that the importance of his works has still not been fully realized. B. B. Polynov's judgment of V. V. Dokuchayev therefore seems very accurate to us, a judgment, incidentally, that has grounds for a broader interpretation. "Great people may be divided into two categories: One is made up of those whose magnitude is subject to the laws of time. Great people of this type are comparatively easy to create: kneeling before them is enough. But the farther they get from us, the smaller they become in our eyes and eventually they vanish on the distant horizon without a trace. The other category of great people is not only not subject to this law, but manifestly contradicts it — as they get farther away, they grow larger in our eyes."[9] Indeed, V. V. Dokuchayev is still not subject to the "laws of time."

The definite unity in the object of study of geography was also perceived by D. N. Anuchin. Although he distinguished individual branches within geography, he also divided it into two parts — *general* and *regional* (specific), stressing that this division is relative and takes place within the whole. "Geography is subdivided naturally into two large sections: *general, earth science,* and *specific,* or *regional geography.* The object of study of the first is the whole earth, its whole surface, whereas that of the second is the individual parts of the surface, countries, and regions. The development of both of these sections are closely related. The greater the number of countries that are thoroughly studied with regard to various geographical questions, the more complete and reliable is the material that general geography can use for its comparisons and conclusions; on the other hand, the more perfect our knowledge is of the forces operating on earth, the forms they change and the phenomena they cause, the better we can understand the phenomena and forms of an individual country and the clearer are its distinctive characteristics.

"The task of regional geography is to compile a complete, accurate, and clear geographic picture of the country being described: its land and waters, the relief of its surface, its climate, the vegetable and animal worlds, and the human population. The description must be based on the present moment, but it must be remembered that the earth's surface is undergoing continual change, that what is visible and exists now is the result of conditions that gradually evolved during the present and previous geological periods. Therefore a proper understanding of a country's surface forms, its landscapes, and phenomena of life can be obtained only investigating its past and studying the processes that caused consistent change."[10] D. N. Anuchin's theoretical views have a common philosophic foundation with the concepts of V. V. Dokuchayev; thus, despite the disparity in the character of the works by these two scientists, we can still see many common elements in their approach to geography as a whole and in their understanding of the true nature of its object of study. Instinctive materialism with elements of dialectics was the common philosophic foundation that shaped their views and that produced the common elements in their understanding of the unity of geography.

Although he adhered to a materialist world view, D. N. Anuchin contended that geography was a science that had a common, specific, material object of study, namely, the natural conditions that evolved on the earth's surface and human society, which is inextricably bound up with these conditions. Nature and man must therefore be studied in terms of their unity and interaction. Thus, D. N. Anuchin also came very close to a correct understanding of geography's object of study and of the essence of the geographic environment.

In considering man within the framework of nature and affirming quite rightly that geography should study not only natural conditions but also social conditions, D. N. Anuchin, understood, as V. V. Dokuchayev did, although somewhat nebulously, the qualitative peculiarties of human society that distinguish man from the rest of nature. This is evident at least from his emphasis on the division of geography into branches, which follows logically from the complex character of its object of study: D. N. Anuchin correctly concluded that the relative differentiation of geography was law-governed and hence inevitable and that geography, by his definition, was a complex of sciences, each of which is capable of development on its own. When he did concrete geographical research, D. N. Anuchin, as a rule, distinguished the study of social subjects from the study of natural subjects. But his recognition of the necessity of differentiation in geography was never carried to a positivist affirmation of the need to liquidate geography as a whole. He was an opponent of a branching-out geography. He always interpreted differences between branches as differences *within a whole*, which was impossible to comprehend by simply summing up the results of specialized study.

Although he correctly understood the need for analytic research, D. N. Anuchin also perceived the need for *combining analysis with synthesis*. For example, in discussing the development of geography, he emphasized that ". . . one should expect, on the one hand, more specialization in the various sections of earth science and, on the other, a closer alliance of the various geographical disciplines."[11] Therefore, we cannot concur with the assertion that D. N. Anuchin regarded physical and economic geography as completely autonomous and disparate scientific disciplines. This point of view, imputed to D. N. Anuchin by A. A. Grigoryev,[12] is unconfirmable.

On the contrary, all of D. N. Anuchin's work contradicts such contentions. Because he understood the need for specialization, D. N. Anuchin also stressed that this specialization must occur *within* geography, without breaking its unity. He said that specialization in geography is not always beneficial, that in a number of instances a synthetic approach has advantages over an analytic one and is absolutely imperative in geographical work of a regional character. "However, such specialization [i.e., specialization in individual branches — V. A.] cannot be applied in the case of specific earth science or so-called regional geography, that is, the synthesis of geographical data concerned with a certain country or part of the world. Here one has to make use of all data that exist in cartography, physical earth science, bio- and anthropo-geography, with the addition of data from ethnography, statistics, industrial-commercial and cultural development, in order to obtain as complete and integrated a picture as possible of the country, its nature, population, culture, its situation and importance among other countries."[13]

Thus, the classical authors of Russian geography, above all V. V. Dokuchayev and D. N. Anuchin, came very close to a correct understanding of the true nature of geography's object of study, although they were unable to formulate their propositions with sufficient clarity.[14] Russian classical geography, in general, including the works of V. V. Dokuchayev and D. N. Anuchin, left us quite a few very fine traditions and a sizable scientific heritage that we are still far from fully assimilating, specifically in geographical theory.

At the beginning of the present chapter we spoke of the overdifferentiation of geography and the one-sided development it spawned. This kind of one-sidedness continues today, and the problem of analysis and synthethis is still one of the most important in geography.

It is well known that by analyzing the parts of a whole, and with regard to geography, by analyzing the individual elements of the geographic environment, which are singled out from the whole and subjected to independent inquiry by individual geographical sciences (geomorphology, climatology, industrial geography, and so forth), a more profound understanding of these elements is achieved. The study of specifics, the anatomization of the whole, is a necessary condition for understanding the whole, and, with regard to geography, for understanding the landscape envelope of the earth. Analysis is a necessary, but not the last, step in the cognition and study of the whole; this could be true only if the whole consisted of the simple sum of its parts. Therefore, the differentiation of a science is merely one aspect of its development and is not capable by itself, without combining with synthesis, of assuring an understanding of the whole.[15]

Yet, as A. M. Ryabchikov quite rightly observed, "some Soviet writers have gotten so carried away with this differentiation that they are even inclined to affirm that geography as a science has already ceased to exist. It is supposedly nothing more than the name of a group of geographical sciences that deal on their own with various aspects of the geographic environment. We hold a different point of view."[16]

Analysis without synthesis, and the impossibility of one person covering all forms of motion, sometimes lead to the loss of the notion of science as being integrated and to a subjectivist, idealistic world outlook instead. "The approach of the mind (of man) to an individual object and the derivation of a replica (i.e., concept) from it *is not* a simple, direct, inert act, but a complex, dichotomous, zigzag act that *includes* the possibility of straying from life toward fantasy and more than that: the possibility of the *transformation* (and an imperceptible transformation, unrealized by man) of the abstract concept or idea into *fantasy* (in the final instance, God)."[17]

The whole is not only the sum of its constituent elements. The earth's landscape envelope therefore cannot be understood solely through its differentiated study. It is necessary to apply synthesis — a method that is opposite to analysis in form but complements it in content.

The one-sided development of physical and economic geography and the widening of the gap between them may be considered, with some reservations, a positive rather than a negative phenomenon for the second half of the nineteenth century and even the beginning of the twentieth century. The separate study of individual components of the natural environment, and of the subfields of population geography and economic geography, was more helpful than the attempts to create integrated pictures of the geographic environment based on geographical determinism or idealistic monism. But now that Marxist philosophy enables us to see the unity of the material world without confusing qualitatively different patterns, now that we understand the *indirect* character of the interactions between society and nature, this kind of gap between the two geographies and the refusal to study the geographic environment as a whole are totally unjustifiable.

It would appear that now the mutual character of the interactions between society and nature should be completely clear to everyone. Nobody is likely now to dispute the proposition that nature's influence on society and the changing of nature by society for its own ends are not two *separate* processes but merely two aspects of a *single* process of interrelations between nature and society in which both sides, that is nature and society, undergo change." . . . In order to produce, people enter into certain ties and relations, and it is only in the framework of these social ties and relations that their relation to nature exists and that production takes place."[18] Consequently, the ties in the dialectical unity and in the interactions between nature and society are carried out *indirectly*, through social relations, without knowledge of which it is impossible to understand these interactions. In this process, human society is the active and leading part of the dialectical unity that it constitutes together with the rest of nature. In the light of this correct proposition, a study of the modern geographic environment that takes up only the complex of its natural components and disregards both the influence of human society on this complex and the effect of social laws, is, to say the least, a reversion to times past in geography.

There is an even more archaic ring to the assertions that the geographic environment as a whole (including its social elements) can be comprehended by a purely natural science — physical geography, whose object of study ostensibly is precisely this geographic environment as a whole.

Attempts to study the nature of the ties between natural and social phenomena, unfortunately, are absent not only in the geography of the past. For example, in a 1957 article by D. A. Armand that contains many interesting and valid ideas, one senses very strongly the author's fear of being accused of confusing social patterns with laws of nature and a fear of being associated with the adherents of a "unified" geography.[19] It is evidently for this reason that Armand obstinately leaves aside the question of human society's influence on nature, of the interaction between them and of the social elements of the geographic environment, although none of this, it would seem, could possibly be skirted in an article with the purpose of defining the subject matter and tasks of physical geography. At the present level of development of geography, to speak of the subject matter of physical geography and, in doing so, to disregard social production, is at least an evasion of the most important theoretical question now facing Soviet geographers.

The development of the individual geographical sciences without a general theory has resulted in a loss of understanding of the common character inherent in the object of study of geography as a whole, and in certain instances it has resulted in departures from geography to related sciences (geology, biology, applied economics). Furthermore, the predominantly analytic development of geography, in turn, increases the gap between its subfields, especially between those whose objects of study develop under fundamentally different laws. This kind of gap is attributable to the abnormally unbalanced trend in the development of geography, a trend that was promoted to a large extent by the reflection of positivist views in geography.

Of course, the significance for the development of geography of theoretical conclusions based on empirical research is extremely great. We contend that here it is *especially* great. In our view, the development of the geographical sciences is inconceivable without the constant addition of new factual material, above all descriptive. To lose touch with factual material and instead merely theorize and methodologize poses a great danger that often leads to glaring errors in the consideration of theoretical scientific questions. Therefore, those scholars who say that "theory should be concrete" are partly right.

But by no means can the role of theory be reduced only to the generalization of factual material. The essence of an object of study can never be understood in the process of direct inquiry alone. Only phenomena can be perceived and investigated directly. In order to proceed from the cognition of phenomena to the cognition of essence, *theory is indispensable*; specialized, theoretical research is needed, and hypotheses are needed. "In theoretical natural science, which unifies its views of nature as much as possible into a single harmonious whole and which even the most feeble-minded empiricist cannot do without, we very often have to operate with imper-

fectly known quantities, and consistency of thought has always had to help still inadequate knowledge to progress
further."[20] "If we wished to wait until material was ready in
pure form for ascertaining a law, this would mean suspending
thoughtful inquiry until then, and for this reason alone we
would never discover the law."[21] And without establishment
of the laws of science, similarly, it cannot possibly develop,
because "the concept of a *law* is *one* of the stages of man's
understanding of the *unity* and *connection*, the interdependence and integrity of the world process."[22] Theory and
hypotheses as approaches to theory should and usually do go
further than factual material permits. Specialized theoretical
works, as we know, can therefore also contain hypotheses
and speculative elements, which, of course, does not mean
that theoretical works should consist of nothing but hypotheses. There need be no fear here that subjective elements will
also find their way into such theoretical research. In the long-
term development of science, these subjective elements (sup-
positions) will either be completely or partially discarded, or
they will be incorporated into theory, thereby taking on an
objective character. In short, theoretical research, like expeditionary works, requires a certain amount of boldness. In the
quest for new discoveries, there should be less fear of error;
overcautiousness, a product of the destructive criticism one
still runs across, leads inevitably to stagnation in theoretical
thought.

The development of geography can no longer proceed in the
direction of analysis alone. In this connection one must regard
as profoundly erroneous the opinion that efforts to compose
synthetic geographical works, for instance, in the form of integrated descriptions of countries, regions, and microregions,
are merely a popularization of the results of specialized geographical research. This opinion is sometimes based on the
fact that when geographers compose synthetic works, they
inevitably make use of conclusions from analytic research in
the specialized geographical (and sometimes not only geographical) sciences. It is on these grounds that synthetic
works are sometimes dismissed as compilations of little scien-

tific merit. The implication is that an article written on the basis of a direct study (analysis) of karstic phenomena on the outskirts of a village called Ivanovka is an inquiry of a scientific character, whereas a work that synthesizes the results of analytic research by many geographers and that furnishes, as a result, a more or less integrated picture of the nature, population, and economy of a country or region turns out to be not scientific, but compilatory or, at best, of little scientific merit (second-rate science of a scholastic kind).

This sort of attitude toward integrated geographical works (physical-geographical, economic-geographical, and regional-geographical) leads to a rejection of geography as a whole and, in effect, leads to a rejection of physical geography, to a rejection of economic geography, and to a rejection of regional geography and earth science, since every broadly geographical work is unavoidably less analytic than synthetic in character and is inevitably based on the results of specialized research. Underestimation of the scientific merit of synthetic works usually appears in our country under the banner of the struggle against superficiality in scientific research. Superficiality needs to be combatted, of course, but it is distressing when this fight quashes any desire in geographers to compose integrated geographical descriptions.

Here many forget what is most essential — the necessity and great importance for science of research dealing with the associations between phenomena, without which no branch of human knowledge can develop. They also forget that it is especially important for geography, in particular, to have studies on the associations between phenomena, for these studies constitute "the very crux of geography, its 'nucleus,' without which the purpose of its existence is lost."[23] Overspecialization sometimes brings geographers into a situation in which they are cutting off the limb they are sitting on.

Essentially repeating the old positivist formula, "Every science is in itself philosophy," geographers who underrate the importance of specialized theoretical works and integrated geographical studies reduce geography to the simple sum of the knowledge obtained from the study of individual elements of the earth's landscape envelope, and thereby seem to be saying: "Every geographical science is in itself geography."[24]

The distinct dominance of empirical research and the underestimation of synthetic works inevitably lead, and in geography have already led, to the dominance of inductive methods of inquiry alone. Yet, ". . . induction and deduction are tied to each other just as necessarily as synthesis and analysis. Instead of one-sidedly extolling one of them to the heavens at the expense of the other, there should be an effort to use each in its place, and this can be achieved only if one does not lose sight of their reciprocal association, their mutual complementarity."[25]

We emphasized above the necessity of developing the individual branches of geography. It is perfectly obvious that without a profound knowledge of the individual components of the geographic environment, it is impossible to understand this environment as a whole; and the time has come to call attention to the necessity of composing codified geographical works and broad cross-specialty generalizations. The further intensification of analysis alone, without synthesizing, leads to a departure from geography. For example, certain scholars studying the geography of industry shift, in effect, to a study of its economics, and from a study of relief as one of the chief components of the geographic environment they shift to a study of it in terms of geology.

This excessive preoccupation with specialized research exists not only in geography, and in every case it is beginning to do great harm. In any realm of human knowledge, specialization is progressive only under the indispensable condition that emphasis of a certain subfield is accompanied by the maintenance of ties with its other subfields. In delving into his own subfield, the specialist remains a geographer only if he sees and understands the general geographical significance of his work. Otherwise he will inevitably turn into a representative of the sciences related to geography.

Our opposition to the one-sided, exclusively analytic development of geography by no means signifies an underestimation of analysis. It is perfectly obvious that synthesis is impossible without previous analysis and geographer-specialists, who are primarily analysts (although there are synthetic aspects to their work as well) are unquestionably making a large and useful contribution. It would be a delusion to interpret our discussion as a call against specialized geographical disciplines, for it is absolutely essential that they continue to emerge and develop. We completely agree with I. S. Shchukin when he writes that "the splintering in the course of its development of a broad scientific discipline into narrower daughter disciplines is an altogether normal and regular process for every developing phenomenon."[26] But this "splintering" ("division" would be better) should not lead to the destruction of geography itself, it should be accompanied by consolidation and synthesis. In the words of the same I. S. Shchukin, who never renounced his monistic view of physical geography, ". . . it is important at the same time that the basic methodology of the mother science not be lost, and that the link to it be sustained in every possible way."[27]

The branching out of geography, which is sometimes represented in our country as a necessary and therefore normal form of the development of science, is actually the consequence of a one-sided development and the absence of a necessary methodological combination of analysis and synthesis in the practice of scientific geographical research.

The economic geographer, of course, must know how to use the data of related sciences (say, economics) for his own purposes, and the geomorphologist cannot perform his research without reference to geology. The establishment of business ties with related sciences is a most essential condition of any scientific inquiry; but these ties must be used by geographers precisely *for their purposes. They should enrich geographical research.* When this does not take place, when ties are established with related sciences at the cost of *ruptured ties* between the geographical sciences, when a specialized object of inquiry ceases to be specialized and turns into an autonomous whole within which an infinity of new objects of study emerge, then we are dealing with a change into a science related to geography. For instance, analysis passes from geography to physics, which happens with climatologists who turn into meteorologists (i.e., from geographers into physicists) or to economics, which sometimes happens with economic geographers who turn into economists.

Departures of this kind from geography bear witness to the existence of a universal link between phenomena, that is, they confirm the validity of the determinist world view and attest to the relativity of any sort of boundaries (intersections) between the individual branches of science. But they also indicate trouble in geography, which is turning into "*the servant of many masters*" while losing its own goals and tasks. Geographers are beginning to duplicate geologists, biologists, physicists, economists, and so forth *without fulfilling* at the same time their *own* tasks and without producing works that characterize the geographic environment and its individual components. Yet knowledge of the geographic environment is extremely necessary from both the scientific and the practical points of view. Therefore, the total elimination of the relative boundaries of geography that separate it from other sciences can hardly be considered a positive fact.

Attention has been directed many times to the broadened and increasingly intensifying relationship between the geographical disciplines and its related sciences; but emphasis has usually been placed only on the positive side of this phenomenon, whereas the negative side has been obscured. For example, I. P. Gerasimov stresses that ". . . the development of scientific geographical knowledge is closely associated with the development of other natural and social sciences. The geographical sciences make wide use for their own purposes [which is precisely what most often does not happen—V. A.] of the achievements of other sciences and are continually enriched with new scientific ideas and methods that arise as a result of such inter-connections. On the other hand, facts or patterns established in geography are widely used in the working out of various general and particular scientific questions of natural science and the social sciences."[28]

Reality, unfortunately, obviously contradicts I. P. Gerasimov's quoted statement. Of course, the development of geography is closely associated with other sciences. But the departure of geographers into related sciences and sheer empiricism in research is turning geography into a set of individual subfields that work out of touch with each other. Thor Heyerdahl made a witty metaphorical observation about this kind of development. "Specialists restrict themselves so as to dig themselves deeper and deeper until they can no longer see each other from their pits. The results are neatly piled up outside. One more specialist is needed, precisely the one that is still missing. He should not follow the others into a pit, but should remain outside and bring all of the various results together."[29]

The lack of synthesis is having a negative effect in all of the sections and branches of geography, but it is perhaps felt most strongly in regional geography, which is inconceivable without the wide application of a synthetic approach. Meanwhile, demands are being made on geography to provide not only specialized works on relief, climate, industry, or transportation but also scientific books from which one could obtain a picture of the geographic environment of a country (or region), that is, its nature, population and economy *as a whole*.

One might mention in passing the very correct idea, stated by N. N. Baranskiy in 1946, affirming the importance of generalizing geographical works, the shortage of which is now felt so acutely. "It should not be forgotten, after all, that the proper object of study in regional-geographical work is a country or a region; as for the individual elements of nature and the individual branches of the economy, these are merely specifics, individual elements that create an overall picture only in their totality. The specific geographical branch disciplines that have evolved in the study of these individual categories — geomorphology, climatology, hydrography, or agricultural geography, industrial geography, transportation geography — have the task of discovering and establishing specific patterns, each in its own field. They are necessary, but they are insufficient. The trouble is not that they exist but that the generalizing disciplines are weak."[30]

The excessive specialization within geography, which to a considerable extent is attributable to the lack of a general theory, has given rise to another negative phenomenon. The

advanced training of geographers, in substantial measure, has
landed in the hands of specialists who have only an indirect
(and sometimes highly remote) relation to geography.
Geologists, biologists, economists (specifically, international
economists, who are gradually achieving dominance in the
economic geography of foreign areas, meteorologists and historians — in short, representatives of the most diverse sciences — are coming to geographical institutions, increasing
even more the excessive differentiation of geography.

Specialists who come into geography from the outside usually have no scientific notion whatsoever about any tie between the individual components of the geographic environment or about the geographic environment itself, do not know
the history of geography and do not understand the nature of
the geographical method. The more the subfields of geography are isolated from geography, the better it will be for
such geographers: No one will deter them from engaging in
their own specialty, which is often not related at all to geography, and it will be all the easier for them gradually to remove *geographical* disciplines from the academic programs of
geography faculties at universities and to strengthen within
the faculties the centrifugal tendencies aimed at turning them
into conglomerates of individual specialized minifaculties that
essentially produce not so much geographers as economists,
hydraulic engineers, meteorologists, biologists, soil scientists,
and so forth. This situation that has been created in the university system has been troubling the geographical community for a relatively long time. The statements made in this
regard by one of the most senior Soviet geographers, V. N.
Sementovsky, are of particularly great interest. "It may be
said that the greater part of the leaders and teachers of the
'geographical branch specialities' are not geographers ....
Because of this, it also becomes very difficult to train cadres
in a qualitative way, so that they can master geographical
methodology and geographical thought."

"It is difficult to achieve a general understanding of the
tasks of geography with such different types of teachers in the
geography faculties."

"We think that the excessive tendency toward specialization, the tendency toward greater autonomy in each specialty, has been brought into geography precisely by these quite numerous cadres, who by chance were led to work in geography faculties. Just because of this it is still difficult to become a geographer ...."[31] "And it is indeed impossible to demand such a radical reorientation from a specialist who has received an education in his narrow field, worked in it for a number of years and only then, not because of an attraction, but because of a 'combination of circumstances,' entered geography. Initially this is done in the form of an invitation to serve subsidiary disciplines. Then, as the structure of the faculty expands, the subsidiary disciplines are made specialized and now define an entire geographical specialization. And the same cadres are already leaders of geographical education with all of its implications. People who worked in the fields of meteorology, hydrology, botany, and so forth, come to geography without changing their orientations, proud of the 'practicality' of their sciences and their rich arsenal of methods of inquiry."[32] "This is precisely why it happens in our faculties that 'the swan (the synoptic meteorologist) strains for the clouds, and the pike (the hydrologist) is drawn to water ....' But the cart is not moving well."[33] It is not surprising that with each decade broadly based geographers in our country are becoming not more numerous but fewer, and literally a handful are dealing with general geographical problems.

This must be pointed out because such phenomena are not accidental. Overspecialization, which in practice is unnecessary for geography and leads to a departure from geographical analysis, and the wide-scale recruitment of related specialists into geographical institutions without a requirement of geographical orientation in their work — all this is a reflection of incorrect views in the training of geographers and in research work. V. N. Sementovsky was therefore absolutely right when he wrote with reference to overspecialization: "This is far from being innocuous. Such a trend abets those who completely reject geography. This leads to the destruction of geography as a science."[34] N. N. Kolosovsky also wrote about this danger. "However, there were also shortcomings that must be corrected in the immediate future. These shortcomings are associated with a certain disdain for the further elaboration of theoretical questions in geography.

"In recent years, for example, university geographers have permitted to continue under their very noses the well-known methodological estrangement between physical and economic geography and the excessive fragmentation of the science into individual specializations, particularly within physical geography. At the same time, the general line of development of geography for the future seems to have been lost. All this, by intensifying the separation of geographical disciplines from each other, is hampering the development of Soviet geography in the proper direction in accordance with the demands of life. The narrowly vocational specialization of knowledge is not a university trend in science."[35]

Overspecialization, which leads to the study of individual elements of the earth's landscape envelope as autonomous wholes, with a lack of understanding of the commonality in the object of study of all the geographical sciences, is unquestionably detrimental to the development of geography and of its individual subfields. Yet very little heed is being given in our country to the voices that warn of this danger.[36]

The division of geography into two juxtaposed sciences (physical geography and economic geography) that has thus evolved at present was, to a significant degree, also the consequence of a singularly empirical approach to the study of individual elements of the earth's landscape envelope.

The dissemination of views denying the unity of geography was aided by the influence of modern bourgeois philosophy, which is unable to rise to an understanding of the unity of the material world. The historical development of physical and economic geography, therefore, could not help but proceed along several different paths.

In the capitalist countries, the trend toward separation of physical and economic geography has become evident, especially in recent decades, which reflects the overall tendency of bourgeois philosophy to turn from determinism to indeterminism. The wall between the natural and social spheres in bourgeois science at present is closely associated with the tendency to deny the objective laws that govern the development of both nature and society. Especially active in the capitalist countries are the advocates of pragmatism, a variety of subjective idealism that is close to positivism. The influence of pragmatism, which has now become all but an official philosophy in the United States, is gradually beginning to penetrate into foreign geography as well. It is expressed in a denial of the causality and universal link among phenomena and in the complete rejection of determinism under the banner of overcoming geographical determinism and the material character of the object(s) of geographic study. They are beginning to attach preeminence to the will of outstanding personalities, which is supposedly capable of controlling the course of events and of shaping the conditions of man's environment; sometimes the influence of the geographic environment on society is denied altogether.

At the same time, one can find in the works of many foreign geographers, including American geographers, correct inferences and conclusions, which are usually generalizations of empirical research. But the views of American geographers on geography are not very coordinated, and obviously there are a number of disagreements among them. A collective work by American geographers, issued in connection with the fiftieth anniversary of the Association of American Geographers, is highly significant in this respect. The authors of the work succeeded in presenting a rather thorough picture of the state and direction of the development of modern geography in the United States. The book discusses in detail all of the individual subfields of geography, including even the medical field and the interpretation of aerial photography. However, the general theory offered in it is extremely inadequate. For instance, regarding the subject matter of geography, as a whole, the book is very nebulous: "Today, as in the past, geography is concerned with the arrangement of things on the face of the earth, and with the associations of things that give character to particular places. Those who face problems involving the factor of location, or involving the examination of conditions peculiar to specific locations, are concerned with geography, just as those who must be concerned about a sequence of events in the past are concerned with history."[37]

Among the general geographical ideas in the book, there are, in our view, two valid ones — the idea of the unity of geography and the idea of the regional method.

In regard to the unity of geography, it says: "Almost all scholars who have thought deeply about the nature of geography agree on the essential unity of the field. The various kinds of duality which have been popular in the past, such as regional as opposed to topical geography, or physical as opposed to human geography, seem to have obscured rather than illuminated the true nature of the discipline. The latter kind of separation between the physical and human aspects continues to hinder the full and balanced development of geography, for it persists in textbooks, in academic organization, and also in the research agencies of government and in the organization of the research councils. This separation seems to have resulted from the nineteenth-century attempt to divide all knowledge into science, meaning natural science, social studies, and humanities. Such a division is intolerable for geographers, for they must deal with man as well as that which is not man (now commonly defined as nature), and the two are intimately intermixed wherever man has been on the earth. Geography, which has to do with places on the earth, simply cannot be made to fit into so arbitrary a classification of knowledge. Actually, there is just one kind of geography."[38]

However, the monistic definition of geography is by no means proven in the book. The book itself consists solely of specialized works. Nothing is said, either, about geography's place among the other sciences or about the existence (let alone the substance) of the fundamental differences between the natural and social branches of geography.

Geographical determinism is the object of intensive criticism in modern American geography, but the criticism is such that one has the impression of an effort to deny determinism altogether. Finally, as the quotation shows, American geographers adhere to a locational point of view in defining the subject matter of geography, thereby negating, in effect, the monistic view of geography that they affirm. It is perfectly obvious that if the arrangement of things on the face of the earth is to be considered the subject matter of a science, then that science can have no unity whatsoever, since the arrangement of various things on the earth occurs in conformity with totally different laws.

The book, *American Geography*, as a whole, demonstrates most graphically how few theoretical propositions there are in geography that have been able to unite American geographers. It attests to the absence in the United States of a detailed theory of geography and to the empiricism of the discipline. But it is encouraging, however, that the editors of the collection criticized some reactionary pseudotheories, geopolitics in particular, and that they hold a monistic view of geography in affirming a definite unity in the subject matter and in method.

People who are not familiar with the history of the development of geographical ideas may acquire the notion that the principal difference between Soviet and American geography is that the Americans hold positions of monism and affirm the unity of geography, whereas Soviet geographers hold the position of dualism and affirm a splintered geography, especially since American geographers frequently emphasize this supposed difference between American and Soviet geography.[39]

What really separates Soviet geography from geography in the United States (and in the other capitalist countries)? What are the real differences of fundamental importance?

In general form, these differences may be reduced to the following.

Most present-day American geographers, as was mentioned above, adhere to a monistic view of geography and regard the regional approach as the methodological basis of geography. They are right in this respect. But the monism of

American geography is missing the most important feature — materialistic dialectics, without which it cannot be theoretically substantiated. And we see that American geographers actually are not substantiating the monistic view of geography in theory, and in the practice of geographical research they are rejecting it to an ever increasing degree. American geography is a science in which intensive differentiation is not being accompanied by synthetic, general geographical works. The monism of American geography is purely *declaratory*.

If one embraces monism without taking a step in the direction of dialectical materialism, one cannot possibly understand and explain correctly the nature of the unity that is inherent in the subject matter of all the geographical sciences as a unity of opposites. Hence the American geographers' failure to understand the qualitative specificity of the subject matter of the geographical sciences. Hence the absence of a conception of economic geography as a special branch of geography with its own subject matter and its own methodological characteristics. Furthermore, the subject matter of geography is defined as an "arrangement," which in itself is not a form of the material world.

The situation is no better with regard to the understanding of the character of the regional method. Although they assert quite rightly that geographers study things and phenomena in their territorial complexes and by discovering differences between locations, American geographers at the same time deny the objective character of these complexes and differences and declare them to be subjective categories whose character and nature supposedly depend entirely on the researcher.

This is how they put it: "Acceptance of the region as objective reality has been increasingly criticized by American geographers, and it is flatly rejected in this book as being incompatible with the position that the region is a device for segregating areal features."[40]

"... This denial of the objective and actual existence of regions," N. N. Baranskiy rightly notes, "deals a very grave blow to the entire regional concept and reduces it almost to naught."[41]

Thus, the monism and regionalism of American geography are not at all related to materialism. This kind of monism inevitably turns into a variety of idealistic monism, which is divorced from the practice of concrete research.

Soviet geography has its own theoretical deficiencies, and we shall discuss them below, but it clearly sees the objective character of the territorial complexes it studies. Regionalization in Soviet geography is one of the principal forms of the geographical method by which objectively existing combinations of elements of the geographic environment are discovered. Regionalization in Soviet geography is not only a method of inquiry but also a method of reorganization; it is pointed toward the future and is used in long-range planning of the national economy. Finally, Soviet geography distinctly shows the difference between the social and natural branches of geography. Soviet geographers have created a theory of physical geography and have a correct grasp of the specific character of economic geography as a social science.

Soviet geographers are also working out a monistic view of geography that is based on the establishment of a common material subject matter for all the geographical sciences, on a common methodology and the practical necessity of synthesis. The monism of Soviet geographers (e.g., N. N. Baranskiy, N. N. Kolosovsky, V. N. Sementovsky, I. A. Vitver, Yu. G. Saushkin, A. M. Ryabchikov, A. I. Solovyov, V. T. Zaichikov, S. N. Ryazantsev, Yu. K. Yefremov, and others) is inextricably connected with materialist dialectics and is based on Marxist-Leninist philosophy.

As for the dualistic distortions that still turn up, they by no means define the character of Soviet geography. The concrete research on the geographic environment, especially of a scientifically applied character (e.g., the qualitative appraisal of lands, applied geomorphology, regional-geographical works, local area studies, etc.), that Soviet geographers conduct has long since broken with dualism. The dualism in Soviet geography is substantially extrinsic in character. It arose mainly as a result of overspecialization and does not have any deep roots in Russian prerevolutionary geography, just as it does not have any such roots in the practical *integrated research* of present-day Soviet geographers.

The lag of synthesis in Soviet geography is its chief deficiency, and this lag was the direct reason for the sustenance and even the limited development of dualism in Soviet geography. True, it is declaratory and has not penetrated (and is probably unable to penetrate) into the practice of integrated geographical research, but the voices that propagandize it in theoretical works continue to sound forth.

The dualistic concept is expounded most thoroughly in a book by I. M. Zabelin, who writes that "a unified geography has become historically obsolete and no longer exists as an autonomous science; it has dissolved into two basic sciences — physical geography and economic geography . . . ."[42] Thus, he flatly rejects the monistic conception of geography, which was supported by all of the major geographers of the past and is being developed by many present-day Soviet geographers. At the same time, he does not deny the existence and development of geography as a unified science in the past. It existed, according to I. M. Zabelin, but in the second half of the nineteenth century it expired, because it was supposedly pulled asunder by the more specialized sciences. "At the same time, the specialized sciences, which had developed rapidly, 'deprived' geography of its own object of inquiry and, in a manner of speaking, took it apart: botany 'took' vegetation for itself: zoology took the animal world; geology, rocks; geomorphology, relief; climatology, climate; oceanography, the ocean; and so forth. The same thing was observable in the sphere of the economic sciences: first statistics, and then the specialized economic sciences 'took away' the national economy from geography, and ethnography and demography took population."[43] Thus, the whole, according to Zabelin, completely dissolved into the parts that constituted this whole, and hence he declares the science that studied the whole to be nonexistent.

As has been mentioned, the monistic conception of geography does not have a substantive theoretical foundation in foreign geography, and Soviet geographers are still only beginning to take a new philosophic approach to it. It is for this reason that dualism has now become current in the theories of certain Soviet and foreign geographers.

Questions arise in this connection. Perhaps all of the outstanding geographers of the past who saw the unity of their science really were mistaken? Perhaps geography has really lost its object of inquiry and does not have its own specific methodology? Perhaps now, in the age of the differentiation of the science, there isn't any and cannot be any geographical synthesis, and the integrated, broadly geographical works, particularly regional-geographical works, are merely vestiges of the past? Perhaps a science, in studying a certain form of motion of matter, disappears as a result of its differentiation? Perhaps the unity of a science can be sustained only as long as qualitatively important differences within the whole that is being studied are not known? In other words, perhaps the cognition of a whole (in this case, the whole that geography studies) ceases with the development of the process of cognition of the individual parts that constitute this whole?

If these questions can be answered in the affirmative, then those who hold a dualistic view of geography will prove to be right. But if the answers are negative, then the scientific untenability of geographical dualism will become obvious, and the basic contours of the monistic theory of Soviet geography can be sketched.

The rest of this book will endeavor to answer these questions.

CHAPTER FIVE

The Landscape Envelope and the Geographic
Environment. The Subject-Matter Essence of
the Unity of Geography. The Influence
of the Geographic Environment on Society.
Determinism in Dialectical Thought.

The observed gulf between physical and economic geography
also has certain objective bases. It would be wrong to attribute them only to the erroneous views that have become current in geography or to its excessive differentiation with inadequate synthesis. This gulf can also be traced to the fundamentally important differences between physical and economic geography.

Physical geography deals with combinations of natural elements of the earth's landscape envelope while taking into account the changes occurring in these combinations under social influences. Therefore, no matter what definition of physical geography one adopts, there is no question that it is a
natural science concerned with nature and its laws.

Economic geography deals with combinations of social
elements of the earth's landscape envelope, expressed primarily in the form of territorial and production complexes. Therefore, no matter what definition of economic geography one
adopts, there is no question that it is a social science concerned with the social phenomena that take shape under the
determining influence of social patterns.

It should be added to this that the territorial combinations
of the earth's landscape envelope that physical geography is
concerned with, as a general rule, do not coincide with the
combinations that economic geography is concerned with,
and this is also an important factor intensifying the difference
between these two branches of geography.

If one considers geography a unified science without seeing
these important differences within it, this will lead to a

mechanical confusion of fundamentally disparate patterns, and ultimately to unscientific inferences and conclusions. *We oppose all efforts to substantiate this kind of unified geography.*

But the existence of differences between physical and economic geography need hardly be proven to anyone. This fact is common knowledge and is universally recognized in our country. However, though we see and understand the nature of the differences between physical and economic geography, we regard as incorrect the conclusion based on these differences that economic geography is completely divorced from physical geography, that these sciences are divided by an impenetrable partition, and the conclusion denying the existence of a system of geographical sciences and of geography as a whole. These differences show only one side of the nature of geography. The other side of its character is that the entire complex variety of sciences that make up geography is concerned with a *single common object of study* and is guided by a common method that is basic to all the geographical sciences.

What kind of common object of study is it that all the geographical sciences — both the natural and the social — are concerned with? Why did the study of this object of inquiry require the emergence and development not of one science but of an entire complex system of sciences? Why doesn't the unity of the object of study of all the geographical sciences preclude fundamentally important differences between physical and economic geography?

The answers to all these questions can be found only if one understands the true nature of the common object of study of the geographical sciences, namely, the landscape envelope of the earth.

The landscape envelope of the earth is the surface of the terrestrial sphere (including the sea and ocean floors) with the hydrosphere and atmosphere. Besides the lithosphere, air masses, waters, soil cover, and biocenoses, the landscape envelope contains an entire complex of social elements, that is, primarily the population with the results of its interaction with the rest of nature. More specifically, the earth's landscape envelope includes territorial complexes of the results of production. The topsoil and vegetation that have been altered as the result of human activity, the altered composition of the atmosphere, man-made structures, and so forth, all remain part of the landscape envelope, as does the population itself, despite all of the distinctive features in its development. "Thus, the landscape sphere, or the landscape envelope of the earth, is the broadest integrated concept of modern geography."[1]

Geography views the earth's landscape envelope as an intricate combination of elements with various qualities, and each successive stage of development or newer form in this envelope contains traces of the properties and qualities of preceding stages, which represent less perfect forms of matter. Living matter, being the highest form of the development of matter, contains the properties and qualities of all the other, lower forms of the material world. The landscape envelope, having an extremely complex composition, differs from the other envelopes of the earth primarily in that life originated and is developing within it. The landscape envelope, therefore, comprises the conditions of the origin and development of life, including its highest form, human society; these conditions then became the *environment* for the development of social life. However, by no means all of the landscape envelope became the environment for human society. For instance, one can hardly speak of the ocean floor as such an environment. One should therefore distinguish between the landscape envelope as a whole and that part of it in which *direct* interaction occurs between human society and the rest of nature, which is customarily called the *geographic environment.*" ... The surface world of the earth, which is the geographic environment for human society, is filled not only with human organisms as such; it has been altered and supplemented by the results of their labor, by their products and structures. The magnitude of these additions is becoming colossal and, in a number of instances, is taking on planetary significance."[2]

The geographic environment is part of the landscape envelope. The closest ties and interactions exist between the geographic environment and the other parts of the landscape envelope of the earth. For example, the peaks of the Himalayan Mountains do not belong to the geographic environment, but by distributing moisture (and not only moisture), over vast territories, they exert a very powerful influence on the geographic environment that has evolved in these territories. Changes in the relief of the ocean floor influence the geographic environment of many land regions. The ice of the polar regions is not a geographic environment, but the water level in the oceans depends on it; if this ice melted, many currently settled territories would be covered with water. In short, when one distinguishes the geographic environment within the landscape envelope, it should not be forgotten that *indirectly,* at least, the *entire* landscape envelope of the earth is that environment. Moreover, in the course of social development larger and larger parts of the landscape envelope are being turned into an immediate environment of the population and, consequently, are being included in the geographic environment.

The difference between the landscape envelope and the geographic environment, therefore, is not that great and perhaps is somewhat nominal. The landscape envelope occupies the entire surface of the earth. The geographic environment occupies the part of this surface that is the immediate environment of social development. But even those parts of the landscape envelope that do not constitute an immediate environment for social life have such a potential and indirectly influence the geographic environment, often determining its individual properties and qualities.

In this work the landscape envelope and the geographic environment are considered equivalent, although we really understand the nature of the differences between them. We are doing this because only the geographic environment is the common object of study of all the geographical sciences, both the natural and the social, which is what gives them a definite unity. It is important to ascertain the nature of the unity of geography, which lies primarily in the common character of the object of study. And this common object of study of all the geographical sciences is, we repeat, not the entire landscape envelope but a part of it — the geographic environment — which is what gives us the key to an understanding of geography as a unified science. Therefore, when we speak in this work of the landscape envelope, we shall mean by this concept, with certain specified exceptions, the part of it that contains social elements.[3]

To us it seems totally wrong to view human society only as an exogenous factor acting on nature. This view is one-sided and hence incorrect: It allows only one side of the interaction between society and nature to be seen. Man is a part of nature and cannot escape it. He not only descended from the animal world, but, after distinguishing himself from it, he has remained and will remain forever linked to it by unbreakable bonds. The qualitative distinctions of human society, therefore, cannot be regarded as some special supranatural properties. By using the material resources that exist on the earth, human society is itself a part of these material resources. "He himself [man — V.A.] is contraposed to the matter of nature as a force of nature."[4] The results of the productive activity of society (i.e., the results of its interaction with nature) are also a part of the conditions and resources of its subsequent productive activity (i.e., they become part of the geographic environment).

The view of human society only as an exogenous factor acting on nature is completely in order in the socioeconomic sciences. But it should not lead to the absolute contraposition of society to the rest of nature or obscure the relativity of this contraposition. The geographer must not forget that human society, despite all of its qualitative peculiarities, is a part of nature in the broad sense of the word, and the contraposition that is usually applied between them is merely an assumed convention that points out the differences within the whole." ... We by no means rule nature as a conqueror rules another people, we do not rule it as someone removed from nature, ... we, on the contrary, belong to it by our flesh, blood, and brain

and are inside of it, ... our entire domination of it consists in
the fact that we, in contrast to all other creatures, know how
to apprehend its laws and apply them correctly."[5]

There is no abstract identity between nature and society.
Society is not a mechanical aggregate, it is not a sum of biological individuals. The life of people is not a simple biological
phenomenon, it is a specific quality that differs from the rest
of nature. But this specific quality does not put society *outside*
of the landscape envelope; by existing and developing *inside*
of it, society belongs to this envelope, constituting its qualitatively distinctive part. There is no gap between society and
nature. The relationship between nature and society is a relationship within a whole, within a dialectical unity that does
not exclude but, on the contrary, implies internal qualitative
differences. "The history of the development of society in one
respect differs substantially from the history of the development of nature. To wit: In nature ... only blind, unconscious
forces act on each other and general laws reveal themselves in
the interaction of these forces .... Conversely, in the history
of society it is people who act, endowed with consciousness,
behaving deliberately or under the influence of passion and
setting themselves certain ends. Here nothing is done without
conscious intent, without a desired end. But as important as
this difference may be ... it does not change in the least the
fact that the course of history is subject to internal general
laws."[6] The operation of these general laws is such that the
development occurring as a result of social influence extends
not only to nature but also to society. In changing nature, man
changes himself. "To exist, man must sustain his organism by
borrowing the substances he needs from *the external nature
around him*. This borrowing presupposes a certain action by
man upon this external nature. But 'in acting upon external
nature, man changes his own nature.' These few words contain the essence of the whole historical theory of Marx, although, taken by themselves, of course, they do not afford a
proper notion and need clarification."[7] ... In acting on the
nature outside of him, man changes his own nature. He develops all of his capacities, and among them the capacity for

'tool making.' *But at any given time the extent of this capacity is determined by the extent of the development of productive forces that has already been achieved* .... Once an instrument of labor becomes an object of production, the very possibility, as well as the greater or lesser degree of perfection, of its manufacture depends entirely on the instruments of labor with which it is made."[8]

Whereas it is, on the one hand, an *exogenous* factor that determines many changes in the rest of nature, human society at the same time remains a *part* of nature, constituting its qualitatively distinctive, highest material form. And, as a part of nature, human society is unavoidably one of the components of the productive forces, that is, of the immediate *conditions* in which its own development takes place. Although it is an exogenous factor that purposefully changes the rest of nature, mankind is at the same time a distinctive component of the geographic conditions in which these changes in nature occur. Human society is one of the sides in the interaction of nature and society, but it is also a *result* of this interaction. Therefore, society can and should be studied not only by itself (i.e., as a whole), which is the common task of the nongeographical social sciences (which study specific laws of social development), but also as one of the most important conditions, as part of the material basis, of the future development of production.

Without making a special study of the laws of social development, but equipped with knowledge of these laws, geographers study society not as a whole but as a part of the whole, as part of the geographic environment. They study not the internal laws that determine the development of society, but society's interactions with the rest of nature as internal laws of the development of the geographic environment.

This completely specific approach to the study of society as a part of a more complex whole is applied only by geography (or rather, its social subfields), which is what distinguishes geography from the specialized social sciences.[9]

The conflict between society (i.e., the social elements of the geographic environment) and nature (i.e., the natural elements of the geographic environment), which is rooted *inside* the geographic environment, is a constant factor in the development of production; it can cease only as a result of the demise of human society. The mode of production determines the specific character of this factor's effect, but cannot eliminate it.

The geographic environment includes the part of the earth's landscape envelope in which social life develops, and consequently it also includes a multitude of elements at a given stage of social development that are *no longer* participating in the process of production or have *not yet* been brought into this process. Neglected farmland, mines and pits, canals, and irrigation systems cease to be productive forces of society, just as the producers themselves that are no longer participating in the production process cease to be such forces. Productive forces, like instruments of labor, are inconceivable without reference to their functions, to the process of labor, to their unity with production relations. A machine that is not utilized in the process of labor is useless. But all of the above-mentioned former elements of productive forces, after ceasing to be such forces, remain in the geographic environment and belong to it.

On the other hand, unused minerals, water resources, lands not used for the economy, and so forth, though they are not instruments of labor and do not take part in the process of production, also belong to the geographic environment. The geographic environment is a broader concept than that of productive forces. Moreover, the geographic environment is not simply a condition of the current process of production but also an inexhaustible material source from which new productive forces and instruments of labor are continually drawn into this process; it is in this sense (and not in the sense of causality) that the geographic environment may be viewed as the material basis of social development. Thus, the geographic environment cannot be identified as a combination of productive forces and instruments of labor, but they cannot be separated from each other or contraposed to each other, either. Productive forces, like instruments of labor, do not exist outside of the geographic environment. The geographic environment exists objectively and independently of productive forces.

The geographic environment is at once a condition and a source of the processes of production, which represents a resolution of the perpetual conflict between society and the rest of nature that is inherent in that same geographic environment.[10]

The entire life of people is, first and foremost, interaction between society and nature. Although the character of this interaction is ultimately determined by social laws (production relations), its concrete manifestations are impossible to comprehend if the phenomena of the laws of nature are disregarded. The opposition of human society to the rest of nature is not only in order, it is imperative. Otherwise it would be impossible to understand the laws of social development and to understand production relations. But this opposition is relative. Also in order, but at the same time relative, is the division of sciences into the natural and the social.

Each of the concrete sciences is a reflection in our minds of certain things and phenomena of the real world. Any science — be it natural or social — is a reflection of the processes occurring in the objectively existing (outside of us and independent of our consciousness and will) world. "The world is the law-governed motion of matter, and our knowledge, being the highest product of nature, is capable only of *reflecting* the operation of these laws."[11]

Another important point is that consciousness, being the product of social development, is at the same time generated by a certain physiological apparatus. Man's consciousness, his psyche, is the consciousness of a living creature that lives in a qualitatively distinctive geographic, primarily social, environment, and this environment acts on man's consciousness by way of physiological mechanisms; hence the definite unity between psychology and physiology, because only a physiologist of higher nervous activity can be an analyst of psychological phenomena. Voices of protest were once heard against this union of psychology with physiology, which was carried out by the great Russian scientists Sechenov and Pavlov; opinions were expressed that this sort of tie with physiology (a natural science) would result in the liquidation of psychology, which some scientists considered a *purely social* science. But now no one can deny any longer that it was thanks to the establishment of its *unity* with the physiology of higher nervous activity that psychology grew into a full-fledged, genuinely materialistic science.

The overriding factor in the development of consciousness, that is, the capacity for purposeful activity, which distinguishes man from animals, is not society as such or nature as such, but rather the process of man's interaction with the rest of nature. The formula, "Labor created man," does not give us the right to conclude that man has a purely social origin (in the sense of his separation from the animal world), because labor is a dual process, encompassing both man's interaction with the rest of nature and the interaction of people among themselves.

All the objects and phenomena of the material world have ties with each other, and the discovery of these relationships bring together all of the areas of human knowledge, including those that appeared for a long time to be developing completely independently of each other. The establishment of these ties is still only beginning, but it is already known to be based on general laws operating in qualitatively different forms of matter. For example, it has become possible to construct models of some biological processes; the functioning of the human intellect has been formalized, which enables highly valuable mechanisms to be made. Cybernetics is starting to find the finest threads of logic that connect totally different areas of human knowledge, which once again confirms the *unity of science.* Different forms of matter, therefore, apart from their specific properties, also have common properties, which make it possible to compare and even join together what at first glance are totally incomparable phenomena. It may be assumed that in the near future we shall be able to use the achievements of cybernetics in geography as well.

One of the basic properties common to all forms of matter is the property of *reflection*. It may quite validly be termed a common property of the entire material world. In the realm of nonliving matter, reflection is of a passive character. A resilient girder that is deformed by a force is passively reflecting an external force. For living matter, reflection is subject to more complex biological laws and, to varying degrees, is active in character. In man, reflection takes on purpose. "Man's consciousness not only reflects the objective world, but also makes it."[12] Nevertheless, the reflection of the objective world by man's consciousness proceeds according to the general laws that also determine the reflection characteristic of less perfect forms of matter. But at the same time it is determined by specific, social laws that are a concrete, more perfect expression of the general laws of reflection.

Thus, the property of reflection of the external world, of external stimuli and influences, is inherent in all matter and all of its forms; "... It is logical to assume that all matter possesses a property essentially akin to sensation, the property of reflection."[13]

Man's consciousness reflects the entire material world around him. The natural sciences are a reflection in the consciousness of the objects and phenomena of nature. The social sciences are a reflection of the phenomena occurring in human society, which is the highest form of the motion of matter. But man's consciousness is also capable of reflecting the process of society's interaction with nature, as well as the environment that changes as a result of this interaction. Hence there may exist (and actually do exist) sciences that are at once both social and natural, or intersectional sciences. These sciences include geography, concerned with the geographic environment, which develops not only under the influence of the forces and laws of nature but also the conscious, purposeful influences of human society.

The complex character of the geographic environment (which may be called humanized nature) also makes the process of its reflection (cognition) extremely complex. It cannot be cognized from the position of a unified geography that fails to see the qualitative distinctions of human society, ignores the effect of social patterns and, in particular, fails to reckon with the determining influence of the mode of production on the character of production. But the geographic environment cannot be cognized, either, from the position of a science that separates human society from the geographic environment and views man only as a supranatural category, as an exogenous factor that acts upon nature by utilizing it. Neither approach — be it from the position of a unified geography that fails to see the qualitative distinctiveness of society compared to the rest of nature, or from the position of a "branched-out" geography, which separates society from nature — can sustain the process of cognition of the geographic environment.

The cognition is possible only if one takes account of the fundamentally important qualitative *differences* between the three realms of nature (nonliving matter, living matter, and human society) and of the effect of both the general and their specific patterns in each of these realms. It is possible, in turn, only if one simultaneously takes account of the dialectical *unity* between all the realms of nature, in which human society is a qualitatively distinctive side of the internal interaction — one of the most important *internal* factors determining the development of the geographic environment. This cognition is possible, finally, only if one recognizes the impossibility of completely separating the natural and social elements of the geographic environment, in spite of the contradictoriness of this unity, that is, if one recognizes the impossibility of a separation between the natural and social branches of an integrated geography. Integrated geographical research can be genuinely scientific and genuinely geographical only when one simultaneously takes account of both the differences and the unity that exist in the geographic environment.

In other words, an understanding of the geographic environment is inconceivable without the prior elimination of dualism in all of its forms. Geography can develop only on the basis of materialistic monism. "Thus, the approach that fundamentally denies the possibility of providing an integrated picture of the world in geographical works and affirms, instead, the inevitability of being contented with two pictures of the world not connected into one system — this approach must be regarded in philosophic terms as a direct defense of philosophic dualism in the view of the world, a defense that is out of place in Soviet science."[14]

At the same time, the capacity to reflect (cognize) the environment and the capacity for purposeful activity should be distinguished from social consciousness. The latter is the highest form of consciousness, expressing the capacity of people to understand the interests of all of human society or of one of its classes. It should be noted here that the social consciousness of the foremost class, which today is the working class, simultaneously expresses the *general interests of mankind*.

"Labor is primarily a process taking place between man and nature, a process in which man, through his own activity, is an intermediary in, regulates and controls the metabolic exchange between man and nature."[15]

"The process of labor ... is ... a universal condition of the metabolic exchange between man and nature, a perpetual natural condition of human life, and for this reason it is not dependent on any form of this life whatever, but, on the contrary, is equally common to all of its social forms .... Just as it is impossible to tell by the taste of wheat who grew it, so it is not apparent from this process of labor, what conditions it is occurring under: under the brutal lash of a slave driver or under the concerned eye of a capitalist ...."[26] The reflection of the process of labor in the consciousness, therefore, must not be confused with social consciousness. Such confusion leads inevitably to a misinterpretation of the essence of sciences as only *class* categories. The total identification of consciousness with social consciousness, and of social consciousness with ideology and superstructure, gives rise to an interpretation of the whole of science, and geography in particular, especially economic geography, as an exclusively ideological, superstructural category. This confusion of kindred but not identical concepts leads to a sharp separation of the social sciences from the natural sciences, to a nihilistic treatment of history, to the impossibility of comprehending the interactions between society and nature, and ultimately to the idealistic contention that the world is unknowable, to a denial of determinism, that is, it leads to *indeterminism*.

In actuality, in the real world, as V. I. Lenin taught, there are no pure isolated phenomena. "That a natural, objective connection exists between the phenomena of the world is unquestionable."[17] Both the natural and the social sciences contain ideological, superstructural elements. But both these groups of sciences, as a whole, reflect actually existing objects and phenomena of the objective material world and under no circumstances may they be considered merely superstructural categories.

The propagators of the ideas of the foremost social class, as a rule, constitute the majority in the foremost ranks of science and culture. But this does not mean that the only ones who can move science forward are scientists who adhere to the positions of the foremost social class. Such an allegation, which gives rise to a nihilistic treatment of bourgeois science (and bourgeois geography in particular) is profoundly erroneous.

History has seen examples in which scientists even with a reactionary ideology were revolutionaries in science. In the transitional periods of the history of mankind such instances are quite regular phenomena.

At first the opinion often arises that the term *geographic environment* itself indicates that it is a condition for the development of something that does not belong to it. Consequently, if the geographic environment is a combination of the conditions for the development of society, then human society cannot possibly be contained in it. This is the usual objection to including social elements (especially population) in the geographic environment. The persuasiveness of such objections is illusory: From the formally logical standpoint, it is impossible to develop in an environment and concurrently to be a constituent part of it. In reality this is precisely the case, although this fundamentally contravenes the formal metaphysical mode of thought. "This mode of thought seems perfectly obvious to us at first glance because it is indigenous to so-called common sense. But man's common sense, a most respectable companion within the four walls of its domicile, undergoes the most astonishing transformations as soon as it ventures into the vast realm of inquiry. The metaphysical way of thinking, although it is valid and even necessary in certain more or less extensive areas, depending on the character of the subject matter, sooner or later reaches the limits beyond which it becomes one-sided, limited, abstract, and entangled in irresolvable contradictions: because, behind individual things, it does not see their reciprocal ties; behind their existence it does not see their appearance and disappearance; it forgets their motion when they are still; it fails to see the forest for the trees."[18]

Incidentally, practical researchers, refusing to yield to philosophy, actually long ago added social elements to the conditions for the development of human society, to the geographic environment. Thus, planning of the development of a given sector of the national economy (or of individual economic regions) usually takes account not only of relief, climate, water, and other natural components but also the degree of settlement (specifically, manpower and its specific character), the number of enterprises, the level and specialization of agriculture and so forth, that is, it takes account of social conditions or, in other words, the social elements of the geographic environment.

Furthermore, it is impossible to name even one concrete manifestation of social development that occurred without the social elements of the geographic environment; this may be abstractly imagined only for the moment of birth of human society, but since it was born, society has always been one of the most important conditions of its own development. The geographic environment is a complex combination of both natural and social conditions that evolved historically and continue to develop on the earth's surface. It should be kept in mind here that production in general exists only as a general, abstract concept. In real life there are specific forms of production that are one of the conditions for the development of other forms of production. Consequently, some forms of production are both combinations of social conditions and a social milieu for other forms of production. For instance, the agriculture of a region, which despite its specificity is a form of production, can be viewed at the same time as an environment for the creation and development, say, of a food or textile industry, which is also a form of production. By the same token, food-industry enterprises may be viewed as a condition for the development of a given agricultural specialization; the existence of a developed agriculture or of developed light industries can be one of the conditions for the creation of certain heavy industries, and so on. It is therefore totally wrong to say that the geographic environment contains only purely natural elements (relief, climate, water, etc.). The geographic environment encompasses the entire combination of material conditions of social development that has evolved on a given territory, including, of course, the population itself with its specific characteristics in culture, national traditions, vocational skills, and so forth.

The development of productive forces is accompanied by a contradictory process of change in the interrelations between society and nature. On the one hand, the *direct* dependence of human society on the geographic environment, especially on the complex of its natural elements, *diminishes*. While gaining ever-increasing knowledge of the laws of nature, man is making use of them for his purposes and, thanks to this, is making himself the lord of nature. Lenin wrote: "... That primitive man received what he needed as a gratuitous gift of nature is a fatuous yarn .... There was no golden age before us, and primitive man was completely oppressed by the difficulty of subsistence, the difficulty of the struggle with nature."[19] We find the same idea in Engels: "The first people to separate themselves from the animal kingdom were, in all essential respects, just as unfree as the animals themselves."[20] But later Engels says: "... Each step forward on the path of culture was a step toward freedom";[21] "... The change from the uniform, hot climate of the original homeland to colder countries, where the year is divided into winter and summer, created new needs — the need for dwellings and clothing for protection against the cold and dampness; it created, in this way, new branches of labor and at the same time new types of activity that distinguished man more and more from animals."[22]

On the other hand, the *indirect* dependence of society on the environment *increases,* instead of diminishing, as productive forces develop. The geographic environment has an aptitude for expansion, and this property has long been utilized in economic activity. Now, literally, with each passing decade, more and more parts and elements of the earth's landscape envelope are being incorporated into the geographic environment; the latter, in turn, is beginning more and more to play the role of productive forces in society, since an increasing number of its elements are being incorporated into the process of production. Many natural resources that man did not need at all in the past have now acquired an unprecedented value by becoming tools and instruments of labor. This process of an increasing indirect dependence on the geographic environment will intensify. "Whereas man, through science and his creative genius, subjugated the forces of nature, they are taking revenge on him by subjecting him, although he uses them, to a real despotism that does not depend on any social organization."[23] Human society cannot free itself from this kind of indirect dependence on nature.

The increase in society's indirect dependence on nature and the intensification of the process of interaction between them have been caused primarily by the growth of population, which requires a continual and commensurate growth of the means of existence, an increase in production.

There is no question that the ties between society and nature within the geographic environment grow rapidly in number as production develops, becoming more and more complex and manifold, since, in addition to the increase in society's indirect dependence on the geographic environment, the entire geographic environment itself is developing to an ever-increasing degree as a result of social influence.

It may be said that the development of society has become one of the leading factors in the development of nature, which, from the moment society's influence on it was established, began to change incomparably more rapidly than before.

But the increase in society's influence on nature, the increase in society's indirect dependence on nature, and the sharp increase in the qualitative difference between society and the rest of nature intensify the integrated character of the geographic environment. It is relevant here to recall Hegel's well-known assertion — every form of organization is stronger the more the functions of the individual parts differ from one another. This assertion runs counter to formal logic, but life has demonstrated its validity many times. It may be seen in the structure of the atom. It also pertains to the geographic environment that has formed on the earth's surface.

At present we are living amidst a nature that has been heavily altered as a result of social influence. We are living not in a primary but in a secondary environment that has often become livable for people in regions where it was unlivable in the past (however, reverse examples may be cited). At the same time, people are living in an environment that contains not only natural elements altered by social influence but also social elements.

It would be incorrect to say that rivers are an element of the geographic environment, but canals and reservoirs are not. Cities and settlements, fields and gardens, roads and canals also belong to the geographic environment, as do forests, steppes, rivers, and mountains. Through its interaction with the rest of nature in the process of production, human society is one of the important factors in the alteration of the geographic environment: It brings into this environment more and more elements, which are impossible to understand or explain without using the data of the social sciences.

The object of study of geography is complex in character, requiring not only a synthetic approach to the combinations of all elements that constitute it but also a profound analysis of each individual element in its association with the others, for without this it is impossible to create a synthetic picture of the geographic environment as a whole. This is what accounts for the development of the individual subfields of geography, each of which is concerned with an individual component or group of components of the geographic environment. But this does not mean that these subfields are absolutely autonomous and are completely divorced from each other. Their autonomy is conditional and relative, since they all have a single common object of study — the geographic environment.

Thus, the geographic environment cannot be regarded as some purely natural category or purely natural environment, as some still mistakenly affirm. The geographic environment is a far more complex unity that is not without many internal contradictions. Each new generation of people makes its changes in this environment while being a qualitatively different element of it. Each new social system readjusts and realters the geographic environment for its purposes. Consequently, as was already pointed out, the results of social development (the operation of social laws), while changing human society, indirectly change the rest of nature as well. It would be a mistake to contend that the natural elements of the geographic environment do not depend at all on the mode of production simply because social patterns do not operate in nature. There is no question that the mode of production influences the purely natural components of the geographic environment (from animals to climate and relief), but it does so through natural laws.

The new discoveries in science and technology create new opportunities for changes in the geographic environment toward maximum satisfaction of social interests. It is therefore totally wrong to consider the geographic environment a result only of the forces of dehumanized nature.

Man "... has not only relocated various types of plants and animals, but has also changed the appearance and climate of his habitat and has changed even the plants and animals themselves to such a degree that the results of his activity can disappear only with the overall extinction of the terrestrial sphere."[24] This proposition by Engels can hardly be disputed.

Even if man, cities, industrial enterprises, roads, and so forth are excluded from the geographic environment, it will still contain living nature, which in its present form is largely the result of social influence. "Animals and plants, which are usually considered products of nature, are actually products of labor, not only from last year, but in their present forms are products of changes that took place through many generations under man's control and by means of human labor.''[25] As a matter of fact, does this apply only to living nature? We can now safely say that soils, the hydrographic network, climate and even relief — all these elements of present-day nature have undergone social influence over a considerable portion of the earth's landscape envelope, and none of them can be understood in terms of pure natural science, isolated from the social sciences, since it is impossible to *understand an effect while ignoring the causes*. To view the world of nature without taking account of the influence on it of human practice is to artificially separate nature from man, to deny the patterns in the interaction between society and nature, and to isolate the natural sciences from the social. Yet we are living in an age in which many geological, chemical, and biological processes occurring on the earth cannot be correctly understood if one disregards the social influence on these processes.

Human society does not simply utilize nature, thereby satisfying its pressing needs. By giving nature a new *quality,* society also changes it, thereby changing the geographic environment in which the process of interaction between society and nature occurs.[26] Consequently, the material basis itself of production changes, and the effect of the geographic environment changes. Finally, in acting upon nature and changing it, human society inevitably changes itself; and the more substantial this change is, the more substantial man's influence on nature is. ''... The most substantial and immediate basis of human thought is precisely *man's changing of nature,* not nature itself, and human reason has developed in accordance with how man learned to change nature.''[27] As we see, Engels, though he stresses the misinterpretation by naturalists of the process of interaction between nature and man (they saw only the influence of nature on man), does not doubt the existence of nature's influence on man.

It is obvious that the qualitative differences between society and nature in no way signify an absence of unity between them. Every substantial change in the life of society will inevitably find its reflection in nature; and changes in nature change the conditions of societal life, thereby indirectly changing society itself.

For example, the discovery of fire substantially altered people's relation to the rest of nature and created completely new conditions for the settlement of mankind over the surface of the earth. Man's domestication of animals and the beginning of the cultivation of plants also led to a new stage — both in the history of mankind and in the history of the development of the entire organic world, the entire geographic environment — to the establishment of a totally new type of relationship between society and the rest of nature, and to the creation of a completely new geography of all living nature, including human society. The discovery and productive utilization of atomic energy, in turn, now opens up completely new possibilities in the establishment of links between society and nature and enables one to speak of the inevitability of new substantial changes both in the life of nature and in the life of society.

The influence of human society on nature increases from epoch to epoch. This is attributable to the fact that the possibilities of changing the natural environment increase with the growth of production. In the process there occur not only *quantitative* changes in the sense that society exerts an influence on a large number of elements of the natural environment but also *qualitative* changes, since with each substantial change in production (in favor of growth) there is also a change in the character of social influence on nature. This occurs because production growth is accompanied by changes in social needs, in the demand for the material resources of nature, and, hence, in the character of the utilization of the entire complex of the natural environment. A change occurs not only *in the strength* of social influence on nature but also in the *forms* of this influence. "The economic and industrial activity of man has become comparable in scale and importance to the processes of nature itself . . . . Man is geochemically remaking the world."[28]

It is not surpirsing that many present-day physical geographers, although they consider the geographic environment to be the purview only of physical geography, are still able to see that its study is fundamental to the common character of physical and economic geography. Thus, K. K. Markov writes: "The enormous importance of the concept of 'geographic environment' is also based on the fact that it underscores the connection between the two main branches of geography — the physical and the economic."[29]

A. M. Ryabchikov comes very close to a correct understanding of the common subject matter of all the geographical sciences when he writes: "In our conception, geography is a science of the earth's landscape envelope, which arose as a result of the interpenetration and joint development of the earth's crust, the troposphere, hydrosphere, and biosphere. When human society arose, the landscape sphere became the geographic environment for it. The vigorous activity of mankind and its powerful material and technical base are inseparable from the geographic environment and constitute one of its most important qualities .... Since mankind itself is inseparable from the landscape sphere and constitutes a part of it, geography should also be concerned with the study of the natural conditions and social patterns of the geographic location of population and its migrations .... The successes of the specialized geographical sciences inescapably require a higher synthesis; and, peering toward the future, we are sure that a 'unitary,' or more accurately an integrated, geography will not lose its scientific aspect."[30]

The unity of geography is derived just as unequivocally by some economic geographers from the character of its object of study, that is, the character of the geographic environment. For instance, Yu. G. Saushkin writes: "The natural environment can be examined only by physcial geography, but the latter can study the geographic environment only in cooperation with a social geographical science — economic geography. Thus, the examination of the geographic environment, in which human society lives and toils, is the task of geography as a whole, the task of the whole system of geographical sciences — both physical geography and economic geography."[31]

A failure to comprehend the synthetic character of geography makes it impossible to compose integrated geographical works. Indeed, how is a geographic portrayal of any territory to be composed if one disregards the ties and interactions that exist between society and nature and evolve differently in each country and region? One cannot be a geographer and reject geography at the same time.

It is primarily by concerning itself with the natural-social territorial complexes that have evolved on the earth and have become a conditon for further societal development that geography differs from related social and natural sciences, which study nature and society with the task of ascertaining the patterns of their internal development. And this is precisely what defines geography's place among the other sciences.

The material character itself of the geographic environment and the intricacy of its composition define both the unity and the differences that exist in geography between its subfields. Geography as a whole encompasses all the sciences that are concerned with individual elements of the earth's landscape envelope, and at the same time it is a synthetic science that seeks to comprehend this envelope as a whole and in terms of its territorial complexes. Physical geography encompasses the sciences concerned with elements of the earth's landscape envelope that develop under the influence of natural laws and is a synthetic science studying the natural territorial complexes of the geographic environment. Economic (more accurately, social) geography encompasses the sciences concerned with elements of the geographic environment whose development is governed by social laws, and it is a synthetic science studying the social territorial complexes of the geographic environment.

Thus, geography as a whole (integrated geography), physical geography, economic geography, and any branch of geography (geomorphology, climatology, industrial geography, and so forth) have their own specialized subjects, which represent certain forms of development of matter but are studied by geographers as elements of the conditions of social development. These forms of matter are not studied by any other sciences as parts of the geographic environment.

The study of a subject (e.g., relief) as a part of the whole means that it "is studied in terms of its interconnections with other components of the natural environment."[32] Hence, "geomorphology is the branch of integrated physical geography that is concerned on a broad geographical basis with an individual component of the natural environment — its relief — and uses the same methods as physical geography, and it can rightfully be included among the geographical sciences. Needless to say, a similar conclusion may be drawn with regard to climatology, hydrology, and botanical geography."[33] The study of a subject as a part of a whole (as a part of the earth's landscape envelope) definitely distinguishes the geographical sciences from the other natural and social sciences, while limiting their content, specifically the level of analysis, which in the geographical sciences must be somewhat different from that in the other concrete sciences, which are concerned with their subjects as a whole. At the same time, geographers not only study their specific subjects as parts of the earth's landscape envelope but also investigate their territorial complexes.[34] The geographic environment is an infinity of territorial complexes; outlining them, pointing out the differences between them, describing and mapping them, and establishing the laws that determine the specificity of each concrete complex — these are the specialized tasks of geography in the most generalized form.

If we take a look at the direction of geographical research, we shall be convinced that this is precisely what geographers concern themselves with most often. Thus, it is in the geographic environment that geography as a whole and each of its branches (be it geomorphology or industrial geography) have their specific and common objects of study.

This common object of study, which is examined in terms of complexes and elements by *all* the geographical sciences, is also the basis for the unification of the geographical sciences into one common *system* and for the production of synthetic, general geographical works. In simpler terms, the *unity* of the object of study inevitably implies the definite *unity* of the science concerned with this object of study.

As we have repeatedly emphasized, geography as a whole must provide in its future development integrated portrayals of the life of nature and society "... in the manner of the classical geography of the ancients, but on the incomparably higher level of modern scientific achievements and of the enormous amount of accumulated factual material. At the present stage of its development, Soviet geography can accomplish this by establishing organic interconnections between the geographical disciplines and the development of new disciplines — intersectional bridges' — on the boundaries of the two now existing, if it becomes necessary.

"The possible unity of the whole must be achieved under such conditions through integration, that is, the combining of data from different disciplines to characterize the whole. However, *integration* is not an end in itself, the end is the *unity* of geography, and that is not the same thing. Integration is the way to achieve unity. Integration that does not lead to this unity is extraneous to geography. At the same time one of the specific characteristics of geography ... is inherent in the establishment of interconnections and integration between natural and social components. Geography does not concern itself with any one type of motion but with their interconnections as a whole on a given territory."[35]

Yet the approach to the whole geographic environment as a complex of elements with different qualities has come under criticism in a host of cases. There is still a widespread point of view that the geographic environment is only a *natural* category, and its cognition is completely feasible from the position of natural science, without any recourse to social sciences. The point of view exists that since the processes occurring in society and in nature are very different, any effort to study the interactions between society and nature results inevitably in unscientific conclusions.

Actually this is not quite true, or rather, not at all true. In real life, besides the purely social and purely natural phenomena and processes, there are many that simultaneously have the properties of both natural and social phenomena or processes. Can one deny, for example, that agricultural operations are simultaneously a social and natural process, and that agricultural output cannot be viewed only as output of a social origin, since it is simultaneously a product of nature? And is it possible to study agriculture by approaching it *only* in terms of the natural sciences or *only* in terms of the social sciences? There is no question that an extremely large role in agriculture belongs to the natural factor, whose effect is simply impossible to grasp by resorting to economic laws alone. "The economic process of reproduction, no matter what its social character, is always intertwined in this area (in farming) with the natural process of reproduction."[36]

Agricultural production, the utilization of minerals, forestry, the construction and use of electric-power installations, and so forth — aren't all these pheonomena and processes really social and natural, don't they occur in nature and aren't they at the same time social in character? Don't they really depend on the mode of production? Don't the laws of nature really operate in them?

How can one ever study agriculture, and agricultural geography, in particular, from the standpoint of the social sciences alone? This is just as impossible as it is to study agriculture only from the standpoint of natural science. The close organic connection between natural and social elements is becoming evident to many representatives of the natural sciences, who are starting to understand more and more clearly that a whole host of phenomena in nature can be understood and explained only with the aid of the social sciences. "From Marx's viewpoint, which was further developed in Lenin's works, soil fertility is not only a natural-science concept but also a socioeconomic one,"[37] wrote W. R. Williams.

It is appropriate here to quote a short passage from a work by the Chinese economic geographer Sun Ching-Chi, who has been working successfully in recent years in the theory of economic geography. He writes: "While sitting in the library at Haiyuntsang[38] and looking at the various objects in the yard, I reflected: Buildings are a product of society, materials are the gifts of nature; trees are planted by man, the seeds are provided by nature; the iron bell is the work of human hands, iron ore is a product of nature, ... whereas man is a social animal, he belongs entirely to society.

"But Marx pointed out that man belongs half to society and half to nature. Our Soviet comrades also view people as 'labor resources.' Finally, I mused that thought in general can be considered a purely social phenomenon! But thought also cannot exist without the human brain . . . and, after separating itself from nature, in the final analysis reverts half to nature and even to me myself. And here I realized that nature and society can be sharply separated only in theory; in practical work they are never separable from each other."[39]

What happens then to the classification of sciences? After all, until now the notion existed that research cannot be done in a sphere in which fundamentally different laws of development operate. This is usually the basis for the opinion that the sciences must all be either natural sciences or social sciences. Can these opinions be considered correct? We think not. From the time human society emerged as a distinctive qualitative category of the material world, three basic types of processes began to develop on the earth: (a) purely natural processes, independent of man, that is, processes of the natural development of the earth's landscape envelope (e.g., the action of solar radiation, volcanic activity); (b) purely social processes, which attend the internal development of human society (production relations); (c) production processes, which are simultaneously social and natural, since they are *directed* by the laws of social development, but at the same time are *based* on the laws of natural development of the geographic *environment* as a principal *condition*; whereas these processes determine the relationships between society and nature, they also depend on these relationships.

It is the existence of such interpenetrations between society and nature and of actual processes that are simultaneously natural and social that determines the *conditional* character of any boundaries between areas of human knowledge and boundaries between the natural and social sciences. Specifically, the boundary between physical and economic geography is conditional and *penetrable*. There is no gap or wall between the social and natural sciences, but there are intersections. Furthermore, sciences that are close to each other inevitably overlap. The intersections between sciences most often grow into a subfield of human knowledge, themselves forming at their boundaries new intersections, new interpenetrations between sciences. These new intersections then grow into new subfields of science, with new conditional boundaries, which then turn into new intersections, and so forth. This is a *form of the development of science.*

Integrated geography, as well as many of its subfields, develop in the intersections. Geography cannot be placed entirely under natural science, just as it cannot be placed entirely in the system of social sciences. But within it there are subfields concerned with elements of the geographic environment whose development is governed chiefly by the laws of nature (physical and biological geography), and there is a group of subfields concerned with elements of the geographic environment whose development is governed by social laws (population geography and the geography of economy, encompassed in our country by economic geography). At the same time, just as between sciences in general, there are no absolute or linear boundaries within the system of geographical sciences, nor is there a conditional boundary such as may usually be observed between related sciences with different objects of study.

It is relatively easy to see the intersectional subfields of geography between its individual branches (the geographical sciences). For example, soil geography, as was pointed out, can quite validly be called an intersection between physical and biological geography (essentially it may be included with equal validity in both). Agricultural geography and a whole host of divisions of population geography can be called intersections between biological and economic geography, and so on. Many individual problems, especially of an applied-science character, are also, in essence, intersections between subfields of geography. For instance, the agricultural appraisal of lands by geographers may be called an intersection between physical, biological, and economic geography, since the successful solution of this problem is possible only on condition that data from all three geographical sciences are synthesized.

As geography develops, it will become increasingly difficult to contain physical geography within natural science and economic geography within the system of social sciences. The interpenetration between the geographical sciences will inevitably increase, since this is required by the character of the (single, common) object of study of all the geographical sciences — the geographic environment.

Of course, physical geography will be able to continue studying purely natural phenomena and objects. First, physical geographers can study parts of the earth's landscape envelope that have not yet become part of the geographic environment. Second, this kind of study can be done with a conscious disregard of human society's influence on nature. We are not implying that the natural components of the earth's landscape envelope should not be distinguished and studied. On the contrary, it is perfectly obvious that in a number of instances precisely this kind of approach is needed for the study of certain components of the landscape envelope and even of certain of its complexes.

At the same time it is no less obvious that this kind of purely natural approach is not capable of ensuring cognition of territorial combinations of natural elements of the geographic environment. One therefore cannot regard as correct the situation in which physical geography is aimed entirely at the study of dehumanized nature alone and is declared to be a science that is able to ensure a comprehensive understanding of the geographical environment from the standpoint of natural science.

We consider it a deficiency of physical geography that the social influences on the earth's landscape envelope is taken into account by physical geographers only in certain, still comparatively rare instances and that more often they seek to restrict their study only to natural things and phenomena. "Wherever possible, we try to reconstruct the natural landscape as it was before man's intervention."[40] Would it not be more correct to study not artificially reconstructed landscapes but those that actually exist at the time of the inquiry? After all, such landscapes that objectively exist on the populated part of the earth's surface are completely impossible to study while totally ignoring the influence of human society.

The study of dehumanized nature and the study only of reconstructed landscapes is losing more and more of its practical purpose. It is turning into a mere auxiliary branch of knowledge that helps us to understand modern landscapes but does not assure cognition of them.

"The history of nature and the history of human society condition one another. Everything that has been said is especially relevant to the new Soviet, communist epoch, in which electric power and technology are acquiring unprecedented strength and are being subordinated to planned outlines of social development. In studying the physical-geographic conditions and resources of Soviet regions, the geographer today cannot avoid taking into account the laws and demands of social development . . . . For the purpose of studying nature in regions whose development is completely agricultural, the geographer must be very familiar with the history of the development of the region, its present character, the principles of modern Soviet agricultural technology, agricultural electric power, and the region's approximate economic assignments for the future . . . . For the purpose of studying nature in regions of industrial development, the physical geographer must know not only the surface resources but also the underground resources exploited in the region. The geographer can be less attentive to biogeographic zonality in such a region than to the geography of geochemical zones associated with mineral deposits. Furthermore, in a number of instances, the network of waterways, topsoil, vegetation, and animals may be completely destroyed by industry, roads and cities, relief may be changed, entire mountains leveled, and landscape zones will have to be studied only in paleogeographical terms."[41]

When physical geography, especially regional, is separated from economic geography in the study of the modern geographic environment, it also loses its practical value since the study on any scale of the natural complex of the geographic environment has practical value, by and large, only if the results of this study are utilized in economic activity. But this utilization, in turn, is possible only if there is a concurrent study (and on the same scale) of the social complex of the geographic environment, that is, an economic-geographical study. In addition, there can and should be general geographical research, whose results will provide, with regard to a certain territory, more or less complete data on the whole complex of natural-social conditions (i.e., the nature, population, and economy).

Geography as a whole, therefore, may be called an *intersectional branch of human knowledge between the natural and social sciences*. The fact that geography, like some other sciences, incidentally, does not fit into the customary classification of sciences shows that the latter has ceased to meet present-day needs. A new and improved classification of sciences must be devised; this is one of the current tasks of philosophy. Soviet philosophers are beginning to take some steps in that direction.[42]

Based on the complicated character of its common object of study, geography as a whole may be called a science of complexes. Therefore, V. P. Semenov-Tyan-Shansky is absolutely right in asserting that geography cannot avoid using data obtained by other sciences, and *different* ones at that,[43] to reach its conclusions; but it should not be concluded from this assertion, of course, that specialized geographical research of a primary character is unnecessary. The unavoidability of operating with an enormous amount of heterogeneous material constitutes both the strength and the weakness of geography. Its strength, because when it uses the data of other sciences, geography is able, by synthesizing them, to make a most thorough and comprehensive examination of the territorial combinations of the earth's landscape envelope, something that is unfeasible for other sciences. Its weakness, because the use of data from other sciences frequently leads to a departure from geography into related sciences and causes the research being done to lose its geographical character.

The impossibility of studying the geographic environment

and its territorial combinations (countries, regions, microregions) in terms of any group of laws with the same qualities may also be attributed to the fact that in nearly every individual instance the specific causes of the origin of a given territorial combination in the geographic environment are highly diversified. Very seldom, and only as an exception, can a specific characteristic of a territory be explained by the effect of laws that belong only to one of the above-mentioned types. It is usually necessary to observe the results of laws with *different qualities*. For example, it is impossible to find on the inhabited portion of the earth a modern landscape that resulted from the action of the forces of nature alone, that is, forces governed only by natural laws. At the same time there cannot be a section of territory on the earth whose characteristics evolved without the influence of the forces of nature, subject to these laws.

The geographer cannot avoid dealing with all three types of laws: (a) laws determining the development of nonliving nature, which are the special purview of the physical and chemical sciences; (b) laws determining the development of living nature, which are the special purview of the biological sciences; (c) and laws determining the development of human society, which are the special purview of the social sciences.[44] Moreover, geography cannot overlook the general laws of the development of the geographic environment as a whole, which links it directly to philosophy.

The geography of modern nature cannot be studied if the social factor is disregarded. By the same token, the geography of population and economy is impossible to study only in terms of social or, even worse, of economic sciences. Notwithstanding all the importance of economic laws, it must be recognized that they alone are not enough to define specific geographic characteristics in population and economy. The distinctive features of a country's historical development and of its nature are of great importance.[45]

The unity of the geographic environment implies the operation of certain *general* laws of its development that are concrete manifestations of the laws of materialist dialectics. One of these laws, for example, is the law of intercausality in the development of the individual elements constituting the geographic environment. This law may be formulated roughly this way: *Any substantive change in one of the elements of the geographic environment is inevitably accompanied by a change in its other elements and a change in the geographic environment as a whole.* In other words, every element of the geographic environment, from relief to human society inclusively, is associated with each other in the most tightly knit fashion. The ties between them may be direct or indirect. For example, changes in relief, in the configuration of a sea coast, in the course of a river, and so forth, inevitably cause certain changes in the life of the population, since in all these instances the conditions of their economic activity change.

The correlative character of the links between the elements of the geographic environment results in the fact that in a number of cases *nonsubstantive* changes in one element can cause *substantive* changes in other elements.

The earth's landscape envelope (and particularly the geographic environment) is not some dull uniformity. Its territorial complexes, landscapes, and their groups vary over the earth's surface. They are extremely diverse, which indicates the magnitude of the differences in the conditions of social life: Each type of landscape has its own individual features, and a higher taxonomic unit is characterized not only by the sum of properties of the landscape it contains but also by distinctive, more general qualities and patterns that require specialized inquiry (here, as everywhere else, the whole is not merely the sum of its parts).

Landscape differences are not only determined by differences in combinations of *natural* elements. The typical landscapes, for instance, in Britain are unlike the landscapes in France not only because these countries have different natural conditions but also because there are specific, historically evolved characteristics in the life and economic activity of their populations. In a number of cases the population's economic activity generates highly substantive landscape features. One can speak, for example, about the industrial landscapes of the Ruhr or Silesia, the French *bocage,* and so forth. A most substantive feature of the geographic environment in certain places are population centers, especially large cities (urban landscapes).

The study of dehumanized (reconstructed) landscapes is beginning to lose meaning, sustaining only its historical interest. Moreover, an approach to landscape science as a purely natural science leads to its transformation into paleography and to the conclusion that landscape science has been liquidated. Thus, I. M. Zabelin has reached the conclusion that the landscapes in city areas have disappeared.[46]

Actually landscapes cannot disappear as a result of human activity, they are only modified. Within the limits of large cities, say Moscow or Paris, there are landscapes that differ from rural landscapes only in the degree and character of man's influence on them.

We have already mentioned that in their practical activity people have long been taking account of social elements in the geographic environment. In any concrete manifestation of production, there is always consideration not of natural conditions as such but of combinations of natural conditions with social conditions. Practice in this question has long since overtaken theory (because of the greater influence of indeterminist distortions in the area of theory). However, if theory should illuminate the way for practice, then practice, in turn, shows the direction in which theoretical research should proceed. This rule is now manifesting itself quite clearly in geography.

Experience in geographical research and in economic activity convincingly demonstrates that it is not the invented category of location but the geographic environment or, in even broader terms, the *landscape envelope of the earth as a whole* (but also fragmented into territorial complexes and into its individual elements) that is the *common* object of study of all the geographical sciences.[47]

The geographical sciences cannot, however, be viewed only as divisions of geography as a whole, since each of them has its own particular object of inquiry and its own specific methodological characteristics. But with regard to the earth's landscape envelope as a whole, all of the individual objects of study of the geographical sciences are specific components. This is why the individual geographical sciences are specific branch sciences with regard to geography as a whole, because they study their specialized objects of study as parts of the common object of study of geography — the earth's landscape envelope. This relative but completely definite unity in the object of study is the chief characteristic that consolidates all the geographical sciences into one system, from geomorphology to industrial geography and that enables us to speak of geography as a whole.

Geography as a whole is an *integrated science,* a *system of sciences,* and the amalgamation of all the geographical sciences into one common system is substantive and not nominal in character. It is based on the *unity* of the object of study (the earth's landscape envelope) and the definitely *common methodology* (the geographical method). This is precisely why geography is not a set of sciences but an integrated science.[48]

Thus, the geographic environment, including not only pure nature but also man with the results of his activity, is a complex of elements with *different qualities.* It is the interaction of these elements of different qualities, as well as the influence of exogenous factors (tectonic activity, solar energy), that cause the formation and further development of the geographic environment.[49] Specific new territorial combinations (complexes) of the geographic environment arise and develop in this process, while old ones disappear, and they are studied by the regional divisions of geography. Human society plays a *special role* in the process: It is the only factor that exercises a conscious, purposeful influence on the development of the geographic environment in the direction of improving its living conditions. This is one of the principal qualitative distinctions of the social group of elements of the geographic environment.

It is perfectly obvious that *a combination of elements developing under the influence of laws of different qualities cannot be cognized by a science that confines itself to a study of the effect of laws of the same qualities.* This is a general law, and breaches of it inevitably lead to incorrect conclusions and inferences.

Synthetic works are possible, therefore, only if there is simultaneous consideration of all the laws of different qualities (i.e., of the three groups) that determine in the process of interaction the formation and further development of the geographic environment of both the earth as a whole and of its individual territories — from continents and countries to micro-landscapes.

In studying the geographic environment and the internal causes of its development (or the interaction of its elements), geographers also investigate the effects on it from external conditions, the exogenous factors of its changes (development).

Thus, the geographic environment is simultaneously a condition and a material source of social development and is studied as *a condition of development* by a specialized integrated science — geography — which, consequently, studies the material sources of social life and the conditions of this life without making a special study of the causes and laws of social development. It is through here that the boundary between geography and the natural and social sciences related to it passes, and this is why the territorial approach is so essential to geography: The material basis of societal development is studied by geography in terms of territorial combinations, and geographers ascertain differences between natural and social complexes.

A decisive role in the orientation of the utilization of the geographic environment belongs to production relations. Without question, the mode of production ultimately determines the character of development of productive forces, and changes in the mode of production result from internal contradictions rooted in human society. Throughout the history of mankind, people's spiritual life, their views and political convictions have been determined by the mode of production of material goods. But it does not follow from all this that the environment is capable of exerting only an accelerating or decelerating influence on the development of production and that it cannot determine anything. Marxism points out that the geographic environment by itself *does not determine* the historical process of social development. But this does not mean that it cannot exert a decisive influence at all on certain aspects of social life, especially in the realm of economic activity.

Having been drawn into the process of production, the geographic environment in a whole host of cases exerts and will continue to exert a *decisive* influence in *indirect form* on the development of a territory's economy. In tundra conditions, the population is engaged in reindeer-breeding and fishing, and not in the growing of tea or citrus crops. The geographic environment as a condition of economic activity, therefore, can determine the economic specialization of individual countries and regions, which, even with a change in the mode of production, does not necessarily have to undergo radical changes. Reindeer-breeding and fishing will retain great importance in the life of the tundra population under communism as well, since these occupations are maximally consonant with the geographic conditions that have evolved there. The task of the organization of production under a planned socialist economy is not at all to develop economic sectors while nihilistically denying the importance of the geographic environment and failing to reckon with it, but is the reverse, to develop these sectors in consonance with the geographic conditions, which in this way should indirectly determine many specific features of the economic geography of countries and regions. In certain specific phenomena, associated with the interaction of society and nature, the geographic environment exerts a decisive influence on the formation of many aspects of the geography of population and economy.

It is wrong to see everywhere the effects of socioeconomic factors alone.[50] Efforts of this sort may lead to totally absurd conclusions, the possibility of which was pointed out by Engels. "... According to the materialist view of history, the determining element in the historical process is *ultimately* production and the reproduction of real life. Neither I nor Marx ever said more. If anyone distorts this proposition to mean that the economic element is the *sole* determining element, he is thereby distorting this contention into an abstract, senseless phrase that says nothing."[51] "No one is likely to succeed, without looking ludicrous, in explaining economically ... the origin of the consonant shift in high German that widened the geographic division formed by the mountain range from the Sudetes to the Taunus into a real breach passing through the whole of Germany."[52] Furthermore: "Marx and I are partly to blame for the fact that young people sometimes attach more importance than they should to the economic side. In refuting our opponents, we have had to emphasize the main principle that they denied, and there was not always enough time, place, and opportunity to do justice to the other elements participating in the interaction."[53]

Certain occurrences and phenomena in the geography of population and economy (not to mention the geography of nature), of course, can not only be slowed down or accelerated but also may be *determined,* admittedly in an indirect but yet decisive manner, by the geographic environment and even its natural complex alone, that is, the natural environment. Thus, the geographic environment, natural conditions (the natural environment), and natural patterns can determine, *through* production, certain occurrences and phenomena in social life. The failure, which still occurs sometimes, to understand this proposition, is one of the reasons for the underestimation of natural and social conditions (i.e., the underestimation of the geographic environment) in economic practice. This in turn leads to an underestimation of the practical value in geographical research and to ignorance of the practical value of geography.

We are accustomed to speaking of the natural environment (often misidentifying it as the geographic environment) as a factor capable of decelerating or accelerating the process of social development. This is valid, of course — the natural environment plays the passive role of conditions, it is static compared to the social environment, which Reclus once called dynamic. The natural environment by itself cannot be the direct cause of changes in social life, it cannot be the direct cause of social development. This proposition is not likely to be disputed. But the speeding up or slowing down of social development can influence its direction. For example, certain countries, most often mountainous ones, may be retarded in their movement from feudalism to capitalism if the geographic environment slows the rate of development of productive forces.[54]

It is clear that the more rapidly developing country can reach socialism earlier than a country in unfavorable geographic conditions will reach capitalism.

The geographic environment therefore can be, on the one hand, a reason for protracted precapitalist attitudes, and on the other, a condition calling for a more developed socialist country to help backward countries that can bypass capitalism and build themselves a socialist system. But there may be a completely different situation, in which a country lagging in its development (because of the unfavorable influence of the geographic factor) can fall into an even worse situation under the influence of another, more developed capitalist country that views the rest of the countries as objects of imperialist aggression and colonial exploitation. Here, as always, the influence of the geographic environment will not be direct but indirect. It will appear in the process of production.

The acceleration or deceleration of social development under the influence of the geographic environment cannot be understood in simplified terms. This is not only a purely quantitative category, since the differences in the rates of societal development, determined to a large extent by the geographic environment, may lead and, in a host of cases, do lead (true, in an indirect form) to a change in the direction of social development. Thus, the occasional disregard in our country of the geographic factor, which supposedly cannot determine anything, also leads to errors in the domain of social needs.

It should be noted that the classics of Marxism-Leninism always attached a great deal of importance to the geographic environment in all aspects of social life. In discussing the specific conditions that facilitated the triumph of the socialist revolution in Russia, V. I. Lenin pointed out ". . . the possibility of enduring a comparatively long civil war, partly thanks to the gigantic size of the country."[55]

The influence of geographic conditions on the life of society is incontestable. It would be totally wrong to see the affirmation of this influence as a manifestation of geographical determinism; the scientific untenability of the latter lies not in its view of the geographic factor's influence as the basic condition of societal development but in its ranking of this influence as the basic cause of development. In other words, the scientific untenability of geographical determinism lies in its mechanistic character. The adherents of geographical determinism held that the cause-and-effect relationship could be such that under certain initial conditions the effect of these conditions could be ascertained. Actually, this is feasible only when the conditions remain unchanged and they are regarded as the simple sum of individual conditions; such instances are rare exceptions. The development of any form of matter proceeds under the influence of complex and diversified factors. There are in life a multiplicity of causes that give rise to the properties and development of a given object of study, and this multiplicity is not a simple sum of causes but a complex, integral whole whose results cannot be derived directly from its causes. Without rejecting determinism as such, Marxist philosophy rejects its mechanistic approach. Causal relations are incontestable, but they cannot be viewed as direct relations.

Determinism is not limited to being a theory of *direct* causal ties, as is sometimes thought. It also includes a study of *indirect* ties, in which causal relationships are comprehensively investigated (in terms of necessity and chance, the essential and inessential, internal and external, etc.). This kind of determinism is *one of the most indispensable facets of dialectical thought*.

Rejection of determinism as such leads to a view of things and phenomena merely as coexisting in space and time and to the impossibility of understanding them as causally interrelated, and this results ultimately in subjectivism. Dialectical-materialistic determinism is the methodological basis of the scientific cognition of the material world. Geographical research, as preponderantly synthetic research, can be based only on a determinist view of the world.

Underestimation of the influence of the geographic environment, the geographic factor, on societal life is identified here as *geographical indeterminism,* or the manifestation of indeterminism in geography. The essence of the unscientific, idealistic view of the world, which in philosophy has been given the name of indeterminism, consists primarily in its denial of the causation of phenomena in nature and society, that is, its complete rejection of determinism in the study of both natural and social things and phenomena.

In the past, indeterminism, allied directly with religion, was organically related to subjective idealism (Berkeley, Hume). In our day, indeterminists in many capitalist countries are coming up with theories postulating that man's will is absolute and that he can remake the material world at his discretion, without reckoning with anything (voluntarism). They attempt to prove the necessity of rule by strong personalities, heroes who are capable of remaking nature and society at their whim. In doing so they sometimes deny the causality and interrelations that objectively exist in the material world and allege that scientific cognition of it as a whole is impossible.

The dissemination of views close to indeterminism in Soviet geography has been promoted by "opportunism" [*konyunkturshchina*], a special pseudoscientific approach that has retarded the elaboration of many theoretical problems. In essence, opportunism is a manifestation of a lack of principle in science, the arbitrary subjectivist adaptation of theoretical propositions, of evaluations of facts and evaluations of scholars of the past by misinterpreted demands of the immediate moment. Opportunism is the absence of any firm convictions, and for this reason it often results in vacillations from one extreme to the other.[56]

Although the theory of geographical determinism has come under thorough criticism, this has not happened to the theory of geographical indeterminism. It is largely for this reason that geographical indeterminism has proved more viable and in geographical theory has proved to be the most widespread of all the erroneous theories. Here is full vindication of the rule that says that *the most dangerous mistakes are those that go unnoticed.*

That is why we are going to return many more times in this work to the indeterminist distortions that appear both in theory and in practice. These reiterations are also necessary because geographical determinism has frequently been criticized in our country from a viewpoint close to that of indeterminism.

The exponents of essentially indeterminist views erect an insurmountable wall between human society and the rest of nature. This is based on the fact that it raises to an absolute the specificity of the laws of social development. The entire interaction between society and nature is reduced to nothing but society's capacity to change nature and to man's utilization of it. History is interpreted only as the simple alternation of human needs that arise and are satisfied. In the process it is forgotten that reality is much more multifaceted and complex, and man, although he has distinguished himself qualitatively from the rest of nature, has remained and will remain forever a part of it; they forget that "... history itself [i.e., the development of human society — V.A.] is an *actual* part of the *history of nature* and of man's molding of nature. As a result, natural science will include the science of man to the same extent as the science of man will include natural science: It will be *one* science."[57]

Failure to understand this reciprocal penetration between the natural and social sciences leads to flagrant indeterminist errors in geography. Moreover, criticism of certain errors based on geographical determinism has sometimes done more harm than good, because the criticism has often denied altogether the scientific value of works by determinist geographers, which in reality is most considerable. The names of some major scholars have begun to be mentioned in our country with the almost exclusive purpose of denying the scientific value of their works, and they are being referred to even as pseudoscholars who have brought geography nothing but harm.[58] Instead of pointing out the determinist geographers' past miscomprehension of the essence of the differences between society and nature, and instead of objectively showing their positive role in the struggle against idealism, the essentially indeterminist criticism has been based on a rejection of the ties that actually exist between society and nature, as a result of which cognition of the earth's landscape envelope was concluded to be impossible, since it develops under the influence of fundamentally different laws (i.e., physical-chemical, biological, and social).

Thus, the determinist geographers saw the *causal* connection among phenomena and the *unity* of the material world, but *failed to understand the qualitative difference* between various categories of that world. On the other hand, modern geographers who make indeterministic errors see the *qualitative* distinctiveness of human society that separates it from the rest of nature; but by raising it *to an absolute,* they *sever the causal interconnection* within the material world, *lose* their grasp of the *unity* between nature and society, and therefore oppose investigation of the things and phenomena that make up the combinations of elements developing under laws of different qualities (natural and social laws).

Geographical indeterminism leads to a rejection of comprehensive investigation of the earth's landscape envelope, since it sharply separates and contraposes human society to the rest of nature. Indeterminism rejects geography as a science, as a definite area of human knowledge with its own specific subject matter and method.

Refutation of the principal idea that has pervaded the entire development of geography, that is, the deterministic idea postulating causality in the origin of geographic phenomena and their interconnections and in the pattern of their territorial distribution, leads inevitably to indeterminism. The particular is always viewed by geography only in its relation to the whole. If there is no relation, there is no geographical inquiry. That is why rejection of determinism is always accompanied by rejection of geography as a science: and rejection of geography, in turn, leads objectively to affirmation of the unintelligibility of the geographic environment in terms of its integrated complexes, since cognition of these complexes by adding up the knowledge of their individual elements is impossible, because the whole is not the simple sum of its parts. Overestimation of natural conditions in society is one of the most important manifestations of geographical determinism, just as their underestimation is one of the most important specific manifestations of geographical indeterminism. It is appropriate here to say once more that in a whole host of cases the natural environment actually plays the *decisive role,* by *determining* the possibilities of specific interrelations between society and nature. Marx pointed out in this regard that "to this day the art of catching fish in waters without fish has not yet been invented."[59] Under any social laws, extracting industries can arise and develop only in places with the appropriate mineral deposits. Agricultural specialization in a number of instances is determined by soil and climatic conditions, and so on.

Man cannot be *absolutely contraposed* to the rest of nature, but this does not rule out the existence of a constant conflict between society and nature. Man's cognition and changing of nature is simultaneously a process of satisfying social needs. Satisfying one set of needs brings forth new needs, and deeper knowledge of nature makes it possible to satisfy these new needs, which in turn leads to new needs, and so forth.

The conflict between society and nature is continually obviated and continually reemerges, without ever disappearing. "Just as the savage, in order to satisfy his needs ... , must struggle against nature, so civilized man must struggle, in all social forms and under all possible modes of production.[60] It is man's struggle against nature, waged in the process of production, that is the chief factor determining the progress of human society. And this struggle will continue for as long as mankind exists.

The geographic environment is characterized by a continual internal process of metabolic exchange between its natural and social elements. It was only as a result of this process that human life became possible. However, the mutual influences between the qualitatively different elements of the geographic environment (social and natural) are not stable. They change, and although a more active role in these changes belongs to human society, it would be erroneous to underrate the importance of the purely natural factors that influence the process of this metabolic change.

Production and the character of its development are ultimately determined by social patterns, but it should not be inferred from this proposition that one can ignore the specific characteristics in natural conditions that are often determined by laws of nature not dependent on the mode of production. Nature's influence on the entire course of production has always been and always will be very great, since this is the matter from which everything is created and on which everything is based. "The concept of economic relations also includes the *geographic basis* on which these relations develop."[61] But along with all that has been said, one must not forget, of course, that the geographic environment cannot determine the development of human society, since it is, as we have already pointed out, a *condition* of this development, and a condition cannot be a *cause* of development. One must not confuse conditions of development and causes of development, as the exponents of geographical determinism did.

It must not be forgotten that although the influence of the geographic environment also includes social elements, it is, with respect to the process of development of human society as a whole, an *external influence* that appears in the process of production. "Materialist dialectics holds that external causes are a condition of changes, and internal causes, a basis of changes, with external causes acting through internal causes.[62] In examining the factors that influence the formation of a phenomenon, it is always very important to see their interaction and to see the importance of the leading factors that are the basis of the changes. It is especially important not to confuse the *causes* of a phenomenon with the *conditions* that favor or impede its emergence and development. For example, the appearance of Petersburg was engendered by the course of Russia's historical development, and its advantageous geographic situation was a condition that promoted both the creation itself and the subsequent development of the city. Geographic situation by itself, therefore, cannot be a cause, (impetus) for the emergence of a phenomenon, although it will always be one of the conditions (favorable or unfavorable) both of its emergence and its subsequent development.

The importance of the geographic situation of a country (or region) for its economic and political life was frequently discussed in the works of the founders of Marxism-Leninism. Engels, for example, wrote in 1848: "... Part of Germany has lagged far behind the level of development of Western Europe. Bourgeois civilization has extended itself along sea coasts and large rivers. On the other hand, lands that lie far from the sea, especially unfertile and rugged mountain areas, have remained a haven of barbarism and feudalism. This barbarism has been concentrated especially in the southern German and southern Slavic countries that are removed from the sea .... The Danube, the Alps, the rocky mountain barriers of Bohemia — these are the bases of the existence of Austrian barbarism and the Austrian monarchy."[63] Marx and Engels, and later Plenkhanov and Lenin, always emphasized that geographic situation, like geographic conditions in general, plays a different role at different stages of social development.

But man's dependence on nature is a perpetual category, because man himself is a part of nature, an element of the geographic environment in which his struggle with nature takes place.

The influence of geographic conditions (the geographic environment) on all societal life is extremely great. In economic practice (activity) this influence is truly difficult to overestimate, and its underestimation always and completely *inevitably* leads to the most grievous consequences, sometimes extending to *natural disasters*. If one takes the natural environment alone, its influence on the development of economy has been, is, and always will be very large. "*Industry* is an *actual* historical relation of nature, and hence of natural science, to man."[64]

If natural conditions and their local specificity are ignored, it is impossible to form a correct understanding of local characteristics in the development of the productive forces of individual countries and regions, and it is impossible in a number of cases to form a correct understanding of the causes of migration by peoples and of the shifting of trade routes, the specific character of the historical process of individual countries and the specific features in the culture of individual peoples. Denial of the natural environment's influence on human society inescapably leads to voluntarism, just as denial of the role of chance inescapably leads to fatalism.

Marx wrote that "... it was the wind that liberated Holland. It made ... land here for the Dutch. As early as 1836 Holland had 12,000 windmills with 6,000 horsepower in operation, which prevented two-thirds of the country from reverting to a swamp."[65] The importance of natural conditions in societal life is so great, and underestimation of this importance so dangerous, that geographical determinism was even given its due by some Marxists, including such a major theorist as G. V. Plekhanov. He somewhat overrated, for example, Mechnikov's book, *Civilization and Great Historical Rivers,* in which no clear-cut distinction was drawn between conditions and causes of development. Plekhanov wrote that "the properties of the geographic environment condition the development of productive forces, the development of productive forces conditions the development of economic and all other social relations."[66] Elsewhere he affirmed that "... the course of events has been continually subordinate here [in Russia — V. A.], as everywhere else, to natural conditions. The relative peculiarity of the Russian historical process, in fact, may be attributed to the relative peculiarity of the geographic environment in which the Russian people have lived and functioned. Its influence has been extremely great."[67] Finally, he asserted that "... the development of productive forces itself is determined by the properties of the geographic environment around people."[68]

These quotations speak for themselves and need no clarification. But Plekhanov understood perfectly well the *indirect* character of the geographic environment's influence on human society. He went far beyond the primitive forms of geographical determinism that we saw in some of the eighteenth-century French enlighteners. The erroneousness of Plekhanov's view on the role of the geographic environment lay merely in his somewhat one-sided approach and in the exaggeration that appeared in his works as soon as he touched on instances of the natural environment's influence on social development. But Plekhanov correctly saw not only the fact itself of the geographic environment's influence on social development, he also understood how this influence is carried out and by means of *what* relations it is manifested in people's lives. He was incorrect only to the extent that he regarded, albeit in indirect form, the properties of the *natural* geographic environment as the determining factor that creates the *social* environment.

It should be recalled that criticism of Plekhanov's deterministic errors have appeared in our country in quite a few works, whereas his indeterministic underestimation of the geographic environment has come under almost no criticism. Moreover, critics of Plekhanov in Soviet literature, especially philosophic literature, have sometimes turned "in the other direction," assailing the geographical determinism in his works from a viewpoint that essentially is close to geographical indeterminism. Finally, they often fail to see the profound and correct ideas in the works by Plekhanov in which he dealt with the questions of the interaction between nature and society. A proper scientific analysis of his works, especially for the purpose of using them to develop theoretical concepts in geography, has yet to be done by anyone. This is without question one of the important tasks confronting Soviet geographers and philosophers.

In the way of a conclusion summarizing the first five chapters of this book, the following may be said:

1. The object of study of all the geographical sciences is a

concrete form of the material world. The subject matter of geography is the landscape (or geographic) envelope (sphere) of the earth.

2. The part of the earth's landscape envelope *within* which human society originated and develops is called the *geographic environment.*

At present there is practically no essential difference between the geographic environment and the landscape envelope, within which the life of human society takes place. Therefore, with the exception of certain specialized divisions, all the geographical sciences, like geography as a whole, have their common object of inquiry precisely in the geographic environment.

3. The geographic environment consists of three groups (complexes) of elements. The development of each of them is governed by their own specific conformity with law. The first group (inorganic) develops under the influence of physical-chemical laws. The development of the second group (organic) is governed by biological laws. The third group (social) develops under the determining influence of social laws.

The geographic environment is a complex, contradictory unity in which a struggle of opposites takes place. This struggle, primarily between complexes of its elements, is the chief force determining the internal causes of the development of the geographic environment as a whole.

4. The laws that govern the development of the inorganic complex of elements of the geographic environment continue to operate both in the group of organic and in the group of social elements. The laws that govern the development of the organic group of elements continue to operate in human society (in the group of social elements).

This fundamentally important fact, which unifies all the elements of the geographic environment, is a most important basis for the constant links between all things and phenomena of the material world. It also allows one to find *properties common to all forms of development of matter.*

However, these laws do not operate in reverse. Social laws do not operate in the biological sphere or in the inorganic complex of the geographic environment. Biological laws, in turn, do not operate in the inorganic complex. Thus, *the more complex and the higher the form of development of matter* in the geographic environment, the *more complex the combination of patterns influencing this development*. At the same time, the higher the form of motion of matter, the greater number of laws of development of the material world it *possesses for its reproduction,* subject to the determining influence of its specific patterns.

5. The complex, contradictory character of the geographic environment defines the complex (integrated) character of the science concerned with it. The group of inorganic elements of the geographical environment is studied by physical geography. The group of organic elements of the geographical environment is studied by biological geography (which is developing in our country within physical geography). The group of social elements of the geographic environment is studied by social (economic) geography.

But the three complexes of elements of the geographic environment are not separated by an impenetrable partition; there are intersections between them, and all of them put together constitute a *unified whole*. Therefore, in addition to the separate study of each complex of elements in the geographic environment, a science is needed that, by generalizing the research of the "three geographies," could comprehend the geographic environment as a whole. The geographic environment is not only the sum of its constituent elements. Hence rejection of geography as an integrated science inevitably *leads to the conclusion that the geographic environment is unknowable*. No matter how many individual elements in the geographic environment are brought forth, no matter how many new geographical sciences arise in this connection, geography will always be indispensable as a science of the geographic environment as a whole. *The differentiation of geography cannot lead to its elimination.*

By consolidating all the sciences concerned with the elements of the geographic environment, geography is simultaneously a system of these sciences and a synthetic, integrated science of the geographic environment as a whole. Although it is the object of study for all the geographical sciences, the geographic environment is also the subject matter of geography, which, by generalizing and synthesizing the results of research by its elements, creates unified theories of it.

This, in our view, is the *subject-matter essence of the unity of geography.*

6. Geography is a science concerned with a structurally complex subject matter. Its differentiation is therefore an inevitable and completely necessary process. Historically this process went from the general to the particular. Arising from observations of separate occurrences, geography originally was not a differentiated science. Later, as it developed, accumulated geographical knowledge, and improved analytically, two branches were distinguished in geography: One concerned with the natural complex of the geographic environment, the other with its social complex. In our country these branches have developed as physical and economic geography.

Subsequently physical geography, in turn, spawned branches concerned with individual elements of the natural complex of the geographic environment, which then became their subject matter. Thus physical geography, while remaining a branch of geography as a whole, was itself transformed into an integrated science and simultaneously into a system of sciences. The same is happening to economic geography, although its process of specialization is slower.

But this process of differentiation of geography is only *one side of the unified process of its development* — the *analytic* side. There is another side — the *synthetic* side. It is manifested in the increasing interpenetration between the geographical sciences, in the establishment of general patterns of development of the elements of the geographic environment and in the effort to create integrated pictures of its territorial complexes.

Such is the historical course of development of geography, conditioned in a law-governed way by the unity of its subject matter and method. It is also conditioned by the unity of science as a whole, which does not recognize internal, insurmountable barriers at all. The unity of science attests to the relativity of any of its division and to the relativity of the classification of individual sciences.

7. The history of geography clearly illustrates its unity. For a long time physical and economic geography developed as one general science. They have a common history and prehistory. In the process of their subsequent differentiation they were reshaped quite naturally into special sciences with their own specific tasks and methods. But at the same time geography as a whole retained the common goals of inquiry that cannot bring its individual divisions and branches (physical and economic geography in particular) to a complete separation. Geography as an integrated science will inevitably continue to develop on the basis of the achievements of all its constituent branches.

The history of geography vividly contradicts the spokesmen for dualism in geography, the exponents of two geographies. It is precisely for this reason that they are forced to reject this history, playing the role of "Ivans who have forgotten their heritage." They aver that geography before the mid-nineteenth century was not a science. The science, in their opinion, took shape only when the process of differentiation of geography occurred. Moreover, efforts are still being made in our country to deprive economic geography even of this abbreviated history.

This attitude toward the history of economic geography is a striking manifestation of nihilism, concealed most often by high-sounding phrases alleging that economic geography could not have existed before the appearance of Marxism. No one can deny that Marxism and the construction of socialism have added and continue to add new content to economic geography. But this proposition by no means applies only to economic geography, and it is simply impossible on this basis to deny the existence of economic geography in the past (and in the fairly remote past).

Seeing only one side of the development of science, the opponents of integrated geography are playing the role of its liquidators, attempting thereby to deny the objective process of the development of science as a combination of analysis and synthesis.

Yet it is precisely now that geography is passing through a pivotal period, in which the predominant analytic trend in its development is no longer sustaining the process of cognition of the geographic environment. The enormous amount of empirical data that has accumulated at present is not receiving the necessary generalization because of the scarcity of synthetic research and the lack of a detailed theory of integrated geography as an integrated science. Therefore the denial of geography's unity is now especially deleterious not only to theory; but by incorrectly orienting scholars, it gives geographical research an unbalanced direction.

8. Familiarity with the history of geography shows that the theories of geographers were always closely associated with various philosophic concepts. The materialistic trend in philosophy was the basis of the deterministic world view in geography. The mechanistic character and inconsistency of pre-Marxian materialism were manifested in geography in the form of geographical determinism. Although it adopts determinism as one of the most necessary aspects of dialectical materialism, Soviet geography discards *geographical determinism*, which in our day has completely demonstrated its scientific untenability and is a basis for the development of various pseudoscientific theories.

The idealistic trend in philosophy was the basis for the indeterministic views of geographers. In every epoch indeterminism *impeded* the development of science, strengthened its *tendency toward exclusive empiricism,* and led to fragmentation, to the establishment of an artificial separation between sciences and to the total opposition of social science to natural science. Indeterminism has always denied the possibility of a monistic world view and has denied the possibility of the development of geography as an integrated science with a common object of study and a single methodological basis.

CHAPTER SIX

The Methodological Essence of the
Unity of Geography. Regionalization as
a Specific Form of the Geographical Method.
The Geographic Division of Labor.
Economic Geography and the Economic
Sciences. The Location of Production and the
Locational Definition of the
Subject Matter of Economic Geography.

Criticism of geographical determinism (Hettner's theory in particular) has frequently been accompanied by a nihilistic underestimation and even denial of chorology, or, more accurately, territoriality. Yet territoriality (chorology) has always been and always will be a mandatory condition of any geographical inquiry.

Science as a whole, as well as all of its divisions, were created by man for man. Science is a *social category*. Hence it studies the material world with a purpose, to satisfy social needs. But this practical purposefulness in the development of science cannot, of course, be interpreted in a simplified way. Concrete scientific research may not have any direct practical value. Cognition of the objective world can proceed without reference to the direct demands of practice and may even go on for many generations without improving the life of humanity. The achievement of a result from a scientific inquiry and its practical utilization may be separated by entire centuries.

At the same time it is indisputable that ultimately the results of the process of cognition are always put to practical use in one form or another. In this connection, one can observe that each branch of human knowledge has its own distinctive approach to its subject matter, an approach that is based on the specific character of inquiry. Geography is concerned with an objectively existing material object of study — the earth's landscape envelope as a *condition,* as the *environment* in which human society lives and develops, and as the *material basis* of social development.[1] Geographers study nature, population, and economy not by themselves (other sciences are concerned with this) but only as the major complexes of the geographic environment, as parts of a whole. The basic task is to reveal specific differences in conditions (in the environment) so that these differences may be subsequently taken into account in practice, in the process of production. Geography investigates the earth's landscape envelope in the process of its development, as expressed in specific territorial complexes.[2] Thus, the task of the geographical sciences includes bringing out spatial differences that give rise to the formation of territorial complexes. From this it is clear that the principle of territoriality (location) expresses not the essence of the subject matter but the specific methodological approach to its study.

The territorial (chorological) approach is the methodological basis of every specific geographical science, because they are all concerned with elements of the geographic environment as parts of a whole and have the task of examining the territorial differences and complexes that actually exist.

The idealism of Hettner's theory was shown primarily in the fact that he affirmed territoriality (chorology) to be the *subject matter* of geography. More specifically, the exponents of chorological theory confused subject matter with method, mistaking the methodological basis of geography for its subject matter. Such confusion sometimes also occurs in the study of history, in which development in time is occasionally identified as the subject matter of history; but the principle of development in time is history's methodological basis and cannot be its subject matter.

Indeed, history is concerned with human society (or nature) as one of the forms of the material world, tracing the course of its development as expressed in time differences. The category of time is not the subject matter of the historical sciences but its methodological basis, just as territoriality (a more specific form of the category of space) is the methodological basis of the geographical sciences. Location and development are not concrete forms of matter, they are abstract concepts that attest only to two aspects of the existence of matter that do not exist without each other.

The natural-historical and geographical methods are two concrete forms of dialectical materialism, each of which takes into account one of the aspects of the existence of matter. And precisely because they take account of inseparable aspects of the existence of matter, both of these methods are inextricably bound up with each other. The territorial approach, which is the methodological basis of the geographical sciences (or the geographical method) is inevitably used in conjunction with the historical method.

The natural-historical method, in turn, is the methodological basis for all the natural sciences and is inevitably used in conjunction with the geographical method. This proposition is common knowledge and is usually not disputed. For this reason those comrades who say that all the geographical sciences are historical sciences of a sort and that history, in turn, is inconceivable without geography are right in a certain sense. But the methodological community of the geographical and historical sciences is *by no means absolute* and does not preclude methodological differences between them. Both the natural-historical and geographical methods are used by history and by geography, but their significance in each of these two sciences is different. In geography, the natural-historical method is of exceptional but, by comparison with the geographical method (chorology), secondary significance; it is applied primarily to understand and explain correctly the territorial differences that have evolved in the landscape envelope and also to make scientific predictions of impending changes in these territorial differences. Without the natural-historical method, geography would be reduced to description.[3] Without the geographical method, geography is inconceivable altogether.

Geography's connection with history is by no means restricted to the realm of methodology. In essence, history, like geography, does not fit into the classification of sciences that is widespread today.

The science of history in its present form is subdivided into the history of nature (natural history) and the history of people (social history). Both of these aspects are inextricably bound up with each other, infuence each other, and constitute a unity, a single, common, integrated science — history. Perhaps one of the most successful general historical works is the one by Zdenek Nejedlý, which depicts the history of Czech nature and the history of the Czech people in terms of intercausal relationships.[4]

In studying nature or society, history has as its basic task the investigation of differences in nature or society, as expressed in the category of time. But in examining the patterns of development and the forms in which this development is expressed in time, history cannot overlook geographic conditions as a territorial category, as the place where this development takes place.[5] Therefore, all the historical sciences also make use of the geographical method (i.e., chorology), which helps historians to comprehend and explain correctly the forms of social life or distinctive characteristics in nature that evolve at a certain time. Without this method, it would be impossible to ascertain local peculiarities in the process of historical development. Without the geographical method, historical inquiry would inevitably suffer from extreme sketchiness, abstractness, and many highly important characteristics would be overlooked or would be inexplicable.

In short, when studying the methods used by a given science, one must see the *principal,* or *basic, method* without which that science is inconceivable, and the *secondary* methods that this science uses in an ancillary capacity. The historical and geographical (in a broader sense, spatial) methods as reflections of the categories of time and space are especially widespread. It is difficult to name a specific science that completely ignores the location of its object of study on the earth (or, in a broader sense, in space). Every science that takes account of the territorial factor inevitably uses the geographical method.[6] This applies primarily to all the concrete natural and economic sciences. As for the historical method, no science can exclude the history of its subject matter from its sphere of study and proceed without the historical method. Geographers clearly cannot study the geographic environment without reference to the historical periods in its development that have been established by the science of natural history.

There is no method whatsoever that is the monopoly of one science. Moreover, the methods devised in one science are often widely used in other areas of human knowledge and sometimes even with great success. This is quite natural. Every method is merely a specific form of a *single* method of scientific cognition, the method of dialectical materialism, just as all the subject matters of the individual sciences are merely specific forms that do not exist apart from the motion of matter, that is, parts of the common and unified subject matter of science as a whole.

The capacity of each scientific discipline to use not one but several methods attests to the unity of science and to the penetrability of the intersections between fields of human knowledge. Can one conclude from the capacity of each science to employ not one but several methods that it is unnecessary to consider methodological differences between individual sciences under their conventional classification? It seems to us that this question can only be answered in the negative. Specialized sciences may be combined into general (or integrated) sciences only when the sciences being combined have a definite unity not only in their object of study but also in the basic method that they use. We concur with S. V. Kalesnik when he denies the possibility of a classification of sciences based on the principle of methodological unity alone.[7] This kind of unity is indeed insufficient for combining a given group of sciences into one system or into one broader science; and it also seems wrong to us to classify sciences based on the single, albeit fundamental, principle of a common object of study. If one and the same science can employ different methods, then, after all, one and the same object of study can come under the purview of different sciences. For instance, population can be the object of study of several sciences: geography, economics, history, biology, medicine, and so forth. If one classifies sciences according to a common object of study, then in this example population geography, demography, the history of population, and a number of other nongeographical sciences would have to make up one common science. But it need hardly be proven that in reality they differ substantially from each other and belong to different scientific systems.

The combination of specialized sciences into a general (integrated) one, for example, the combination of geographical sciences into geography, is quite natural and actually takes place because all of the sciences being combined form a definite community both in their object of study and in their methodological basis. In geography, this community is based on the landscape envelope, which is the object of study for all of the geographical sciences, and on methodological unity, since every geographical science studies it subject matter in developmental terms, as expressed in territorial complexes. This proposition enables us to speak of their common methodological basis or of a common chorological, or more accurately, geographical method that is completely indispensable to all the geographical sciences. Only this kind of community of object of study and method can be used as the basis of a classification of sciences. A classification of sciences based only on a common object of study or only on a common method indicate a failure to understand the essence of these sciences, which leads to arbitrariness in their arrangement. Sciences may be classified only according to *the combination of objects of study and method that is specific to each of them.*

Geography is a concrete science. It is concerned with a material object of study consisting of various elements that are in dialectical unity. Therefore, cognition of this object of study is possible only if it is approached as a series of interrelated forms of motion that are investigated by examining territorial differences. If this is so, then the underestimation of the territorial approach in the study of the geographic environment and its elements is a profound delusion. Hence, the special importance of mapping for geography. The mapping of the earth's landscape envelope is one of the most important specific manifestations of the geographical method; nongeographical cartography does not exist, although not even map-making is the monopoly of geography.

Cognition of the earth's landscape envelope through examination of its territorial complexes, as has already been mentioned, is inseparable from the natural-historical method. Thanks to the inseparability of the geographical and historical methods, geography is able to portray the past, to depict the present, and to cast an eye toward the future. It would be wrong to limit the tasks of geography to the study and description of the landscape envelope only in its contemporary (for a given generation of geographers) form.

Moreover, even when investigating the contemporary geographic environment, geographers cannot escape acquainting themselves to some extent with its past. Otherwise it is difficult to understand the present. This kind of association of geography with the sciences of the history of the earth and of the history of society has been discussed by M. V. Lomonosov. "And, first, it must be firmly remembered that the visible corporeal things on the earth and the whole world were not at their origin in the state which we now perceive, but, rather, great changes have occurred as is shown by history and ancient geography combined with present geography, and by the changes in the earth's surface that are occurring in our age."[8] The necessity of studying the geographic environment in the process of its historical development and the necessity of comparing its present state with the past so as to find out what changes occurred in it and what changes may occur in the future have been pointed out in various ways by every reputable geographer. For example, Carl Ritter was completely right in emphasizing the necessity of the historical method in geographical research when he wrote: "But not even geography can do without the historical element if it wants to be a true science of terrestrial spatial relations and not an abstract, pale image of localities or a brief guide that provides only a frame and a graticule for a vast landscape, but not thoroughgoing accounts of its substantive relations and its internal and external laws.''[9]

A denial of the methodological ties (methodological interpenetration) between geography and history inevitably results in the separation of space (territory) from time, leading geography to chorology and history to chronology, that is, to the replacement of a material object of study with methodological characteristics.

The application (or nonapplication) of the geographical method is one of the most important criteria for determining whether a given work is geographical. Geographical works always offer more or less concrete theories. They give descriptions of the natural and social conditions that evolved over a certain period on a given territory.

The results of geographical research always provide a more or less complete and necessarily *concrete* picture of certain territorial complexes in the geographic environment or of its elements. These results always can and must be *cartographically located*. A map, as N. N. Baranskiy often says, is one of the most important criteria of whether an inquiry is geographical (although geographical inquiry, of course, cannot be reduced to maps alone).

It may be observed, incidentally, that the results of the study of components of nature and of certain aspects of an economy are often plotted on *general geographical* maps, which once again confirms the existence of a definite unity between all the geographical sciences — both natural and social.[10]

The results of specialized geographical work are characterizations of individual elements of the geographic environment, in close association with the rest of its elements and with the examination of the territorial differences that the element under study objectively possesses.

Physical-geographical study should produce an integrated picture of all the natural elements of the geographic environment or, in other words, a picture of the entire complex of natural conditions in terms of its territorial differences. Economic-geographical study should produce a portrayal of the entire complex of social conditions in terms of their territorial differences. Finally, general geographical study should produce a representation of the whole complex of the earth's landscape envelope, again in terms of its territorial differences. Furthermore, one can take as the object of study either the earth's entire surface, or individual regions, all the way down to micro-landscapes, depending on the degree of detail of the inquiry. This proposition is completely applicable to any geographical inquiry.

No matter what objects (or phenomena) geographers deal with, they are inevitably studied in terms of their intercausal relationships with other objects and phenomena, because they are studied not as entities within themselves but as parts of a whole, as parts of the earth's landscape envelope. Moreover, geography studies and shows the ties and interactions that exist between territorial combinations of the landscape envelope (countries, regions, microregions), primarily from the standpoint of the influence of these relationships on the formation and development of these combinations.

All the geographical sciences, therefore, have a single, common combination, specific to them alone, of an *object of study and a basic method that is indispensable to all of them*. This is what places geography in a special area of human knowledge.

However it would be wrong to think that the existence of a method common to all the geographical sciences precludes differences in the methods of inquiry employed by the geographical sciences. The methods employed by physical geography are substantially different from the methods used by economic geography. Within physical and economic geography, similarly, there are subfields with their own distinct subject matter and method of inquiry. This existence of a distinct subject matter and a distinct method defines the qualitative character of the division of geography and, incidentally, of other broad sciences as well. It is the specialized subject matter and method and the particulars of each subfield of geography that make them sciences capable of relative autonomy. This is quite natural, since "... in order to comprehend certain aspects (particulars), we must separate them from their natural or historical context and examine each one separately, in terms of its properties, its distinctive causes and effects, and so forth."[11] The removal from a whole (for example, from the earth's landscape envelope) of particulars (that is, its elements) for more intensified study is an altogether legitimate and necessary phenomenon. It would be wrong, in the pursuit of geographical synthesis, to oppose the development of individual geographical sciences. What is unfortunate is not that individual subfields of geography are developing but that their growth diverts to itself almost all scientific efforts and thereby hampers the development of geography as a whole. It is an especially distressing situation when one-sided, specialized development leaves few general geographers who are prominent and well-rounded enough to coordinate specialized research.

The problem of regionalization is closely related to questions of the common basic method of geography. In essence, *regionalization* (like map-making) is a more specific expression of the geographical method and is therefore characteristic of all the geographical sciences. Regionalization consists in examining objective territorial complexes of natural and social conditions, that is, territorial complexes of the earth's landscape envelope. It is perfectly clear that examination of these complexes means investigation of objective regions, in which the degree of detail of the inquiry or, more accurately, the *scale of inquiry* will determine the degree of detail of regionalization. As a specific manifestation of the geographical method, regionalization in geography does not have to have great diversity. Geographers can investigate the territorial complexes of an individual component of the geographic environment, leading to specialized regionalization, which brings out, for example, geomorphological regions, climatic regions, botanical regions, and so forth. Physical- and economic-geographical study inevitably produces integrated regionalization, that is, it uncovers physical- or economic-geographic regions.[12] There may also be intermediate forms of regionalization, which reveal the territorial complexes of individual groups of components of the geographic environment (e.g., soil and vegetation zones, industrial and transport regions).

Physical geography investigates territorial complexes in natural conditions. Economic geography examines territorial complexes in social conditions. The laws that determine the formation of natural conditions are *fundamentally different* from the laws that determine the formation of social conditions. The fundamentally different character of these laws generates a greater dynamism in the formation of social conditions compared with the natural environment. The boundaries of economic-geographic regions are therefore less stable than those of physical-geographic regions.

Physical-geographic boundaries are distinguished by relatively high stability, but they also are impermanent, especially now, owing to the intensified social influence on nature. As for economic-geographic boundaries, they are not only less stable but also more conditional than natural boundaries. Intersectional territories of a kind often form between them. It is therefore not surprising that only in comparatively rare instances do economic-geographic boundaries coincide with natural boundaries, and even coincidences of that kind are most frequently an ephemeral phenomenon.

The determination of economic-geographic regions is highly complex largely because of the dynamism of their boundaries and the extreme complexity of the series of causes that form them; such determination should rely on the data of economic geography but also make use of the data of physical geography and the technical sciences. Without the latter data it is impossible to take account of the material and technical base of the economy and the degree of technical possibilties of influencing nature, which are increasing with unprecedented speed in our age of atomic energy.

The importance of the task of determining objective economic-geographic regions lies not only in the fact that it is a basic methodological technique of economic geography for understanding territorial production complexes, but also in the fact that economic-geographic regionalization should lay the foundation for state regionalization, which itself is a highly important method of planning and the national economy.[13]

It is primarily for this reason that Soviet economic geography should be based on the *regional* method of inquiry, making it possible to use the data of economic geography on a wide scale for state economic planning, without which correct long-range planning of our country's national economy is impossible.

Economic-geographic regionalization, as a *method* of cognition of the social complexes of the earth's landscape envelope, is therefore related in the most direct way to state economic regionalization. By distinguishing objective economic-geographic regions, the most important criteria of which are production specialization and integrated development, economic geography lays a scientific foundation for state regionalization.

Thus it would be wrong to consider regionalization by economic geographers completely identical to regionalization by state agencies for the purposes of better planning and management of the national economy. A difference does exist. It lies in the fact that state regionalization is above all the delineation of major economic-geographic regions.[14] When it begins to take on a more fractionalized form, it usually turns into an administrative-economic division of the country's territory and thereby deviates from economic-geographic division.

Furthermore, state regionalization must inevitably take account of the influence of many additional factors (above all political) which, from the standpoint of economic geography, are not regional determinants. For this reason, the boundaries of regions studied by economic geographers do not necessarily coincide precisely with the boundaries of regions defined by state institutions.[15]

Economic-geographic regions are formed under the basic laws of economic development, in which major economic-geographic regions are territorial complexes whose specific character is defined almost entirely by economic patterns.

Therefore, major economic-geographic regions usually combine territories with different, and sometimes sharply different, geographic conditions. In a number of cases this difference in geographic conditions is even one of the factors contributing to the formation of a major economic-geographic region. In major economic-geographic regions we always observe the great internal heterogeneity of the geographic environment.

A somewhat different situation is created in more fractionalized economic-geographic regionalization. Here, too, of course, economic laws exert a guiding influence on the formation of regions. But the boundaries of medium-size (meso-) and small (micro-) regions often coincide with natural boundaries, and their geographic environment does not have substantial territorial differences, since the formation itself of such regions is based most often on a certain geographic homogeneity. The influence of local characteristics as regional determinants is important here to an incomparably greater degree than in the formation of major regions. In a number of instances the general patterns of the economic development of the country as a whole are manifested here indirectly, through the utilization of local characteristics in natural conditions and resources. Economic-geographic microregions, for example, never have a sharply varying geographic environment. On the contrary, they are bounded most often by natural-economic rather than economic borders.[16] In short, the basic economic laws that determine the development of the country's economy as a whole affect the formation of small economic-geographic regions less categorically.

As we mentioned earlier, although economic-geographic regionalization does not yet coincide completely with state economic regionalization and even less so with administrative regionalization, it is closely related to these other types. It can be used on a very wide scale by state institutions as a basis for determining territorial differences in the country's economy.

In a socialist economy, economic-geographic regions (of different levels) expressing a territorial form of economic organization, can and should gradually coincide with administrative-economic territorial units. This is brought about by the activity of administrative agencies, whose functions in the economic sphere are expanding and will continue to expand. It is interesting to note that some socialist countries (Czechoslovakia and Bulgaria, for example) have already achieved nearly complete concurrence between economic-geographic regionalization and administrative division.

The first scientific foundations of regionalization were laid by K. I. Arsenyev (1789–1865), the most prominent Russian economic geographer of the first half of the nineteenth century. A good many interesting ideas on questions of regionalization were advanced by the revolutionary and democrat, N. P. Ogarev (1813–77), who called attention to the determining influence of the laws of political economy on the formation of territorial complexes of social conditions. Like Arsenyev, Ogarev held that economic-geographic regions were spaces distinguished by a clear-cut distinctiveness of natural and economic conditions, but he also attached importance in each space to the leading productive forces. Russian geographers continued to develop scientific principles of regionalization without interruption after Arsenyev and Ogarev. It suffices to recall the works of P. P. Semenov-Tyan-Shansky, who made a great contribution both to the theory and to the practice of Russia's regionalization in the new post-reform socioeconomic conditions. Karl Marx, as we know, showed great interest in the economic-geographic works of P. P. Semenov-Tyan-Shansky, including his works on regionalization.[17]

V. I. Lenin gave a great deal of attention to the problem of economic regionalization and made wide use of the regional method of inquiry in his works. He was the first not to limit himself to consideration only of distinctive characteristics in productive forces, but pointed out the whole importance of production relations as one of the chief regional determinants. He established that under the capitalistic mode of production economic-geographic regions take shape in the process of capitalist development of national economies and of their largest units, in the process of market formation. It was only after the works of V. I. Lenin that economic geographers began to delineate and study regions as production complexes, interpreting production as a combination of productive forces and production relations. V. I. Lenin also developed the method of regionalizing countries according to the character of the mode of production. V. I. Lenin's economic regionalization of Russia correctly showed the basic characteristics of the social division of labor that was proceeding in the country. The theoretical principles of regionalization developed by V. I. Lenin were adopted by Soviet economists and economic geographers, and somewhat later by many scholars in other countries. Lenin's works have become the theoretical basis of Marxist economic geography.

In the new socioeconomic conditions that resulted in the Soviet Union from the Great October Socialist Revolution, economic-geographic regionalization found extremely wide use. It may be recalled that V. I. Lenin always attached a great deal of importance to the correct delineation of economic-geographic regions, both regions that existed objectively and those that would evolve over the long run in connection with state planning of the economy. The drawing up of the State Plan for the Electrification of Russia (GOELRO) was accompanied by the first outlines of long-range regionalization. This plan, ratified by the Eighth Congress of Soviets in 1920, not only provided for the electrification in the formation of individual major economic regions and contained a scientific analysis of the development of the productive forces of these regions on a new electric-power base, that is, it viewed electrification as a regional determinant.

Later the State Planning Committee and a Special Commission of the All-Russian Central Executive Committee, under the chairmanship of M. I. Kalinin, worked out the basic methodological principles for the long-range regionalization of our country. Based on these principles, the whole development of our national economy was invariably related to the problem of regionalization, with consideration of local and natural conditions.

At the congresses of the Communist Party of the Soviet Union (CPSU) attention was focused repeatedly on questions of regionalization and consideration of local conditions. For example, at the Twelfth Party Congress a decision was adopted "On Regionalization,"[18] and at the Fifteenth Congress the Directives for the First Five-Year Plan set the task of "completing the regionalization of the whole country,"[19] so that the entire development of the national economy could proceed with consideration of regional peculiarities. The Sixteenth Congress took note of the great value of the regionalization that was under way, and the decisions of the Seventeenth Congress on the second five-year plan developed an entire program for the location of productive forces, the development of economic regions, and the shift of industry toward the East. The Eighteenth Congress adopted a program of new construction and, in connection with it, plans for the development of economic regions. The resolution of the Congress said: ". . . The Congress believes that in *the location of new construction* in the third five-year plan among regions of the USSR, it is necessary to proceed with the purpose of bringing industry closer to the sources of raw materials and regions of consumption so as to eliminate irrational shipments over excessive distances and to give further development to previously economically backward regions of the USSR."[20] Following the Eighteenth Party Congress, the USSR Academy of Sciences was assigned to give a scientific basis to the economic regionalization of the USSR. The Nineteenth Congress called attention to the necessity of improving the geographic location of industrial enterprises,[21] and the Twentieth Congress called attention to the necessity of taking greater account of local interregional differences in economic and natural conditions and to the elimination of the pernicious effects of excessive centralism and bureaucratism in the management of the national economy. The Directives of the Twentieth Congress set the task of locating agricultural sectors and carrying out region-by-region agricultural specialization, with reference to the natural and economic conditions of each region, and within regions, and with reference to local features in the natural-economic conditions of each collective and state farm.

The decisions of the Twenty-first CPSU Congress are of very considerable and direct importance for the further development of Soviet geography. The creation of a material and technical base for Communism that is capable of providing an abundance of means of production and consumption requires diversified development of our country's productive forces at an unprecedented pace, above all in the eastern regions of the Soviet Union, where the greater portion of natural riches is concentrated. N. S. Khrushchev's report, "On the Plan Figures for the Development of the USSR National Economy in 1959–1965," demonstrated the necessity of large-scale unified construction and set the task of economizing on labor and gaining as much time as possible both in construction and in the utilization of existing enterprises. N. S. Khruschev's report also posed the task of achieving greater geographic differentiation in the building and management of the economy, so as to maximize the use of existing conditions in each region of our geographically heterogeneous country. "In each republic the economic branches that are undergoing the most propitious development are those for which the most favorable natural and economic conditions exist there . . . ."[22]

However this requirement can be fulfilled only if people in applied work are better equipped with knowledge of the specific geographic conditions of the republics and provinces in which they work. Hence the pressing need for better organization and greater scope in geographical research. The solution of the problem of the location of production, which requires better and fuller integration of productive forces in the process of their development, is inconceivable without ample knowledge of the environment in which the process of production is occurring. At the same time, the development of the national economy requires radical improvement of long- and short-range planning and their organic conformity with specific natural and economic conditions. Underestimation or, even worse, disregard of territorial differences in the geographic environment has always led and inevitably will lead to stereotyped planning, which has an extremely harmful effect on the development of the economy.

The decisions of the Twenty-first Party Congress, which repeatedly underscored the importance of geographic differentiation in the fulfillment of the tasks of the seven-year plan, point to the necessity, in effect, of not only economic but also *geographic* planning of the future development of the national economy. Planning, especially long-range planning, can no longer be based only on technical economic calculations. It must make use, on a much wider scale than it has until now, of geographical data — from both economic and physical geography.

Motivated by the instructions of the Communist Party, many prominent Soviet scholars have written works on the problem of economic regionalization. Those who have examined economic-geographic regions and the prospects of their development most successfully are G. M. Krzhizhanovsky, I. G. Aleksandrov, N. N. Kolosovsky, L. L. Nikitin, and N. N. Baranskiy; they have done much for the development of the theory of economic regionalization and for its practical guidance.

As an objective reality, economic-geographic regions also take shape in capitalist countries, where it is also possible to delineate and study them. But the formation of regions in capitalist countries proceeds in an unplanned manner, so regionalization there is feasible almost exclusively for cognitive purposes. In the conditions of the capitalist mode of production, regionalization reveals and shows the results of the unplanned development of productive forces. Regionalization merely registers the effect of uncontrolled laws in territorial combinations of productive forces.

The problem of economic-geographic regionalization in socialist countries is taking on new content. Here it is inseparable from long-range planning. "As we see, the Soviet Union's economic planning is not only of *cognitive,* factual significance, it also envisages the goal of *transforming* the country's economy. It is therefore very closely related to *long-range plans* for the development of the economy and should anticipate the future ten to fifteen years in advance."[23]

All this makes regionalization one of the important factors in the further development of the productive forces of socialist countries. The creation in socialist countries of a system of regional production complexes, tied to each other primarily by their *common territory,* is a form of socialist economy. "It is precisely this form of utilization of productive forces that will give our economy an additional superiority in the economic competition with the highest forms of the capitalist system."[24]

In socialist countries economic-geographic regions are interrelated. Here the relation between the specialization of economic regions and the geographic (territorial) division of labor is a relation between the process of formation of regions and the development of productive forces that leads to it. This is precisely the objective basis of the specialization of regions. A definite, mutually influenced unity takes shape between the specialization of regions and the character of productive forces, which are linked by the geographic division of labor as a form of production relations. But this unity does not, of course, rule out the possibility or, rather, the inevitability of contradictions, since absolute consonance between productive forces and production relations is impossible. Contradictions arise when the relatively correct relationship between productive forces and the historically established division between regions is upset in the process of production. Disturbances of this kind occur because the rate of development of productive forces is usually more rapid than changes in production relations. The rate of development of productive forces accelerates especially when it is aided by the geographic factor, by favorable geographic conditions. Productive forces (especially production technology) develop more rapidly than changes in the interregional division of labor, which is what upsets the necessary proportions between them. This disturbance of the necessary proportions between the interrelated elements of production causes irregularities in its development and begins to impede the further growth of productive forces. This usually leads in particular to an increase in irrational shipments.

It must not be forgotten that the geographic division of labor is the most sensitive aspect of production relations and reacts quickly to changes in the level of technology. The geographic division of labor therefore changes unceasingly, even when forms of ownership and relations between social classes and groups have not changed. Technical progress and changes in production technology find their expression in the reorganization of relations between producers.

But the entire technical (as well as organizational) aspect of relations between producers, of course, is determined by the mode of production and forms of ownership. Ignoring this fact inevitably leads to a revision of the basic tenets of Marxism.

Under conditions of socialist production, economic geography along with economics are capable of forseeing the emergence of contradictions between the interrelated elements of production and therefore can help to eliminate them. The practical role of economic geography here lies chiefly in ascertaining the specific possibilities of utilizing and taking account of local characteristics in the geographic environment to remove conflicts. It is precisely the correct estimation and utilization of local conditions and resources that make it possible to correct the division of labor, which is the specialized purview of economics. Here, undoubtedly, is one of the interpenetrations between geography and economics (which, as we shall show below, does not eliminate the differences in subject matter between these related sciences) and between productive forces and production relations.

It is perfectly obvious that this indissolubility results in the fact that when economic geographers study territorial complexes of productive forces, they cannot (and should not) ignore other aspects of production relations that often have a decisive influence on the formation of territorial combinations of productive relations (i.e., as we mentioned earlier, it is impossible to understand the *effect* while ignoring the *causes*). For geographers in particular, especially economic geographers, it is extremely important to study the social and geographic division of labor, which often has a powerful influence on the economic specialization of countries and regions. Hence, the necessity of also studying production relations, both within countries and regions and between them.

The question of the nature of the geographic division of labor has still not been theoretically elaborated to the proper degree. The current opinion that the geographic division of labor is synonymous with the social division of labor, but only in its spatial expression, cannot, in our view, be considered at all correct. Every division of labor, of course, is social in character; but far from every social division of labor may be called geographical.

It is perfectly obvious, for example, that the division of labor within an enterprise will be technical and economic, not geographic, except when individual parts of the enterprise are located in different geographic conditions. The division between physical and mental labor cannot be called geographic either, even if its evolution varied from country to country. The same may be said about the vocational division of labor.

It is obvious that in a class-oriented society two basic forms of social division of labor should be singled out: the form generated by the class stratification of society and the form generated by the interaction between human society and nature. Under the domination of class-oriented social relations, these two forms often intertwine and even coincide, with the dominant form clearly being the social division of labor generated by the class stratification of society. This form *distorts* and *perverts* the social division of labor generated by the interaction between society and nature.

The geographic division of labor cannot occur in pure form in a class-oriented society, since the relations between it and nature are determined by relations between people based on exploitation of man by man. The differentiation that develops between countries and regions with a capitalist mode of production is determined by relations between antagonistic classes.

In the epoch of imperialism, the process of geographic specialization is subordinated to the interests of the major monopolies of the dominant imperialist states, which develop specializations advantageous to them in the weaker, dependent countries. For the peoples of the dependent countries, this geographic division of labor usually takes the form of colonial oppression, which often leads to hunger. A striking example of this is India before its liberation from colonial dependence.

The social division of labor generated by the class stratification of society, which may be called the *class* division of labor, is a category that arose historically under certain conditions of the class-oriented society and is disappearing in a society where class stratification is eliminated. In the class division of labor, its results (products) are *materialized forms* of social and class relations that cannot have *substantial* territorial differences. It was these relations that Marx was speaking of when he equated the division of labor with private property: "... The division of labor and private property are identical expressions ...."[25] The class division of labor in itself is not geographic in character, and it cannot be called geographic, although in its concrete manifestation, as a rule, it intertwines with the geographic division of labor and appears in geographic form, since it is social relations that determine the character of people's relations with nature.

The process of development of the geographic division of labor in a classless society is totally different. The geographic division of labor in itself, considered in pure form, does not depend on the mode of production. It arose at the dawn of human history and can disappear only with human society. In the geographic division of labor, its results (products) are *materialized forms* of the interaction between human society and the rest of nature. But the interaction between society and nature always bears a pronounced geographic character; it will vary from country to country and region to region under any social structure, since it depends on the geographic environment, which cannot be uniform over the entire surface of our planet.

Based on the territorial differences in the process of interaction between society and nature, the geographic division of labor engenders production specialization in individual countries and regions and hence production ties between them. This economic specialization and these production ties will not only not disappear in communist society, they will develop and become many times stronger.

In a classless society the geographic division of labor appears in pure form. It is freed from the deforming influence of the class division of labor and begins to express the *interrelations between collectives of free producers and nature.* The new production relations that are established in a classless society will not be in conflict with the relations between society and nature. The character of products (production specialization) will be determined not by the interests of the ruling classes but by the interests of the entire population, which inevitably gives rise to conformity between economic specialization and the distinctive characteristics of the geographic environment.

In addition to the geographic division of labor, the *technical-economic* (and the closely related vocational) division of labor will also exist and develop under Communism, based not on man's interaction with nature but on intrasocietal interaction. In addition to geographic differences and patterns, economic laws that determine the organization of production and direct the process of the struggle against nature toward a certain goal will also operate in future society.

Communist society will have its own interactions between forms of social division of labor. But they will not be of an antagonistic character; the effect of social relations on man's relations with nature will be fundamentally different from the effect in a class-oriented society. The geographic environment will cease to be disfigured, and geographic resources will cease to be plundered in the interest of profits. Social influence on nature will be directed toward its utilization but at the same time toward its conservation and transformation in the interests of mankind.

Thus, geography must inevitably include among its interests many questions related to the study of production relations, although they are the specialized subject matter of the socioeconomic, and not the geographical, sciences. This is also necessary because productive forces are not only influenced by production relations, but they, in turn, exert their own powerful influence on production relations, often determining their territorial differences.

A concrete study of our geographic environment shows that social conditions in real life, including the geographic division of labor, are heavily dependent on natural conditions, and natural conditions form under the increasing influence of society. A concrete study of the inhabited part of the earth's surface shows that the natural and social conditions that have evolved on it are interrelated and exert mutual influence, that is, there exists an environment whose constituent elements make up a unity. At the same time, the geographic environment in the course of its development has formed a complex mosaic of differences on the earth's surface. Therefore, in addition to physical- and economic-geographic regionalization, it is possible to ascertain and study objective territorial units with a definite community in nature, economy, and in the socioeconomic life of the population, that is, complexes of the geographic environment as a whole. These territorial complexes are outlined by special boundaries that most accurately may be called historical-geographical, because they arose in the process of the historical development of human society, in connection with the geographic division of labor, and took on a most distinct geographical significance, since they separate territories with different geographic conditions. These historical-geographical boundaries sometimes coincide with physical-geographic frontiers, sometimes with economic boundaries. Although these concurrences are quite regular in each specific instance, they are still insufficient for one to speak of them as some regular phenomenon, since they do not always occur.

The largest territories defined by historical-geographical boundaries are those of states. Smaller territorial units are often (but not always) bounded by the borders of states that no longer exist; sometimes these are borders of past colonization or borders of a settlement by a certain people.

In general, state and national boundaries, if they are not an episodic phenomenon, *always* take on the significance of historical-geographical boundaries, and the territories they circumscribe will *always* have significant differences. If, for instance, a state boundary is passed "over a living body," that is, if it divides a territory that has basically uniform geographic-environmental conditions, then with the passage of time the divided parts of this geographic whole will begin to acquire *substantial* differences in geographic environment, and not only in its social elements, where these differences will be of the greatest degree, but also in natural elements.

This is attributable to the fact that in each state the interaction between society and nature will proceed somewhat differently, and, hence, differences will also arise in the geographic environment, which to a large extent is a result of this interaction. The formation of these differences in the geographic environment will proceed with particular rapidity if a state boundary divides countries with different modes of production.

Conversely, when territories of *different* characteristics are united into a single state, the differences in the geographic environment of the united territory will become less salient (although, of course, they will never disappear completely): they will take on the properties of *internal* differences, since the similarities will be intensified.

For this reason geography cannot limit itself to delineating and studying physical-geographic and economic-geographic regions alone. It must study (and, in fact, does study) territories defined by state, administrative, and historical-geographical boundaries, that is, groups of countries, countries, and regions. In studying countries and regions, geographers came closest to producing general geographical works that describe the geographic environment not only in terms of components but also as a whole, in terms of the organic connection between all of its constituent elements. It is the branches of geography engaged in the delineation and synthetic study of territorial complexes and regions that we put under the common heading of *regional geography*.

Having established the common subject matter and methodology of all the geographical sciences, it is comparatively easy to prove the untenability of attempts to categorize economic geography as an economic science.

This untenability becomes obvious when the object of study of geography is juxtaposed with the object of study of economics (in the broad sense of the word). It is well known that the subject matter of the economic sciences is production relations. "Political economy deals not at all with 'production,' but with the social relations of people in production and the social system of production."[26] Political economy is the basis for all the economic sciences and is their theoretical foundation. "It is precisely on this basis that autonomous economic sciences appear, studying individual aspects of economic life with their specific characteristics in the service of society. At the same time, political economy, being the basic economic science concerned with production relations as a whole, furnishes a theoretical basis for the whole totality of economic sciences. This is essentially what distinguishes the subject matter of political economy from that of the other economic sciences."[27] Political economy is concerned with production relations as a whole, while the individual economic sciences are concerned with their details and with the objective regularities that reflect the process of economic development of a given branch of the economy. It should not be forgotten here that each branch of the economy contains regularities that are not at all different from the basic, general regularities of political economy, although each of these branches also has its own specific characteristics.

The economic sciences are specifically concerned with social (economic) relations in terms of their reciprocal ties with productive forces. They investigate the relations within human society, which is studied as a whole. The geographic environment is therefore viewed quite naturally by economics as an external environment that is contraposed to human society.

Geography is not specifically concerned with social relations. Human society and, therefore, the laws of its development do not come under the range of questions that geography deals with. It is concerned with the earth's landscape envelope and the geographic environment as a part of the landscape envelope. Human society, in our opinion, should be studied by geographers as a part of this whole, as a part of the geographic environment. Geographers, therefore, must study the social conditions (the social milieu) of societal development.

In contrast to the economic sciences, economic geography is not concerned with social relations, either. Its subject matter is the complex and individual elements of the geographic environment that are simultaneously productive forces for society. The differences between economic geography and economics are thus very substantial — they are based on *subject matter*.

However, studying productive forces in their relationship with production relations, economic geography is linked to the economic sciences. It is closely associated with them, since it is related to the system of economic scientific disciplines. The economic sciences, in turn, cannot ignore productive forces, which often influence production relations. Economic geography's close ties with the economic sciences are incontestable.

So we repeat once more that the subject matter of economic geography is the social elements of the geographic environment (population and economy). Economic geography is therefore unquestionably social in character and, to a large extent (but not exclusively), is based on general regularities established by political economy; if these regularities are not taken into account, cognition of the social elements of the geographic environment is totally impossible This is a significant difference between economic geography and physical geography.

But the social elements of the geographic environment are an integral part of the geographic environment as a whole, organically connected with its natural elements. Therefore, the social science of economic geography belongs to the same system as a number of natural sciences that are also concerned with individual components of the geographic environment. Consequently, economic geography is linked to the other geographical sciences by a *common object of study*.

Economic geography does not have this kind of connection with the economic sciences, or with other related sciences, for example, the technical sciences.[28]

The social elements of the geographic environment are territorial combinations of productive forces (including the instruments of labor) in their connection with production relations; they are territorial production complexes, and not production relations, which are studied by the economic disciplines. Therefore, political economy *alone*, despite all of its importance to economic geography, cannot be its theoretical basis. Political economy is not concerned with productive forces,[29] because under a uniform mode of production the productive relationships studied by political economy do not have *substantial* territorial differences, since the common mode of production evens them out to a very considerable degree.

The situation is quite different with productive forces (or, rather, with combinations of productive forces and instruments of labor). Under a uniform mode of production they may have substantial territorial differences, usually in each region and certainly in each country, which always contain their own economically significant distinctive characteristics, although the common mode of production does give a certain unity to all the countries and regions in which it prevails. Therefore, one cannot speak, for example, of *complete* distinctiveness with regard to the historical process in a country (or region). Feudal attitudes prevailed in all feudal countries, giving them common features. Capitalist attitudes prevail in all the capitalist countries, also according them certain similarities. The influence of the geographic environment, of course, cannot turn a feudal country into a capitalist one or a capitalist country into a socialist one.

The existence of casual ties between the development of society and the geographic environment does not in itself create any definite directions in social development. Society develops in accordance with its own inherent laws. Geographic conditions can vary in significance, and even be of diametrically opposite significance, at various stages of the history of mankind. With each step forward mankind seems to reappraise the geographic environment from scratch.

But the development of productive forces is not governed by the mode of production alone. The laws of nature by no means disappear here but continue to exert influence, contrary to the notions of some economists.[30] The degree of utilization of the laws of nature in the interests of human society increases with each new and more perfect mode of production.

The leading and guiding role of the mode of production in the development of productive forces does not eliminate other factors, and in a host of cases even intensifies their effect. The effect of many laws of nature in the sphere of productive forces increases noticeably as the mode of production is perfected and as production grows generally. Therefore, the mode of production does not remove the substantial differences in productive forces, since the effect of *natural* laws on them in different countries and regions can vary *substantially*, which is what causes various types of *interaction between society and nature*, a *uniform* mode of production notwithstanding.[31] Failure to understand this proposition (or perhaps law) leads to theoretical justification of stereotyped methods (especiallly in the planning of the national economy), to a denial of the significance of specific geographic characteristics in individual countries and regions, to underestimation of natural conditions, and to a denial of the importance of taking geographic distinctiveness into account for the economy, that is, it leads to indeterminist distortions.

Of course, in terms of its main, fundamental features, the development in time and the location of production (i.e., development in space) depend on the mode of production, on the basic economic law of the particular social system (feudal, capitalist, socialist). This is why a completely valid distinction is drawn in our country between the economic geography of the socialist countries and the economic geography of the capitalist countries, although this subdivision does not do away with economic geography as an integrated and relatively autonomous science.

But the uniformity of the mode of production and the effect of the basic economic law of a social system cannot eliminate the relative distinctiveness and peculiarity of the development of individual countries. Differences are not even eliminated in the development of individual regions within a country, regardless of what mode of production prevails in it. Feudal China was different from feudal England in many very important respects, and feudal England in many ways was unlike feudal Russia.[32] A common mode of production increases the *similarity* between countries, but can never destroy the *differences* between them.

This again is attributable mainly to the fact that a community of production relations (feudal, capitalist, or socialist) does not erase the differences in productive forces, which form under the influence of the geographic environment, which always varies from country to country and region to region. The effect of the same social laws in different types of geographic environment produces somewhat diverse results, which is what makes territorial differences in public production inevitable; it is totally essential to know these differences in order to direct the development of production properly. Production in *different* countries and *different* regions under a *uniform* mode of production will always have its peculiarities, its local specificity.

By the same token, differences in production lead inevitably to differences in the process of historical development, which therefore also has its own specificity in each country. This specificity, in turn, accords distinctive characteristics to productive forces. But even this is not yet all. Idiosyncracies in production and specific characteristics in the historical process also arise from differences in the external influences exerted on individual countries and regions, influences that it is totally essential to take into account. ". . . The more peculiar the course of our social development became in comparison with that in Western Europe, the less peculiar it was with regard to the course of development in the Eastern countries, and vice versa."[33] In reality there cannot be two countries that undergo an absolutely identical influence from their neighbors. It has always varied; and this cannot fail to make an imprint and intensify the geographic distinctiveness in the productive forces of countries and regions.

Thus, the productive forces of each country (or region), apart from the mode of production, also develop under the influence of: (a) the geographic environment, which exerts an especially strong influence on the growth rate of productive forces and on their specific specialization; (b) specific characteristics in the process of historical development; (c) external factors, that is, influences exerted on a country's productive forces by other countries.

As a result, the productive forces in each country (or region) always have many distinctive, *specific* features, the cognition of which is one of the basic tasks of geography, especially economic geography. At the same time, it is perfectly obvious that attempts at investigating the specificity of productive forces as expressed in territorial complexes will be doomed to failure if we limit ourselves from the outset to the economic approach alone — if we study this specificity in terms of economic determinism. The cognition of geographic phenomena requires a broader monistic approach.

Yu. G. Saushkin came very close to a correct definition of the differences between economic geography and economics when he wrote: "Economists and economic geographers are members of kindred but different sciences. Economists are more interested in the character of production relations between people, the productivity of their labor, the value of their produce and what components it is made up of, and how production technology is distributed. Economic geographers are more interested in *what* is produced and *where*, why a certain *specialization* of production takes shape in a given locality, what the boundaries of the region with this specialization are, how this region exchanges output with other regions and countries, and what natural conditions and resources are utilized for production."[34]

To identify economic geography as an economic science and to base it totally on political economy alone would be, to our mind, a serious error that would cause it inevitably to lose its scientific value.

Yet opinions were turning up in the periodical press until recently to the effect that economic geography should be concerned not with territorial production complexes as a part of the geographic environment but with production as such. It is sometimes even asserted that economic geography should take the place of the economic sciences in making specialized studies of production relations. For example, one may find definitions such as the following: "The object of study of economic geography is production, and not productive forces; and its subject matter is the location of production, and not territorial combinations or complexes of productive forces and not geographic differences in their regional combinations."[35] "... In other words, Marxist-Leninist economic geography is intended to ascertain the role of production relations in the location of production and in the evolution of the territorial division of labor."[36]

"... It is perfectly obvious that physical geography is a natural science and economic geography is an economic science."[37]

All these and similar definitions of economic geography fail to show what its subject matter is. They express the efforts of economists to include in their science a scientific discipline that cannot possibly fit into that field.

A common mode of production cannot, as was mentioned earlier, eliminate local peculiarities, which will always affect productive forces. This, in turn, points to the impossibility of comprehending territorial production complexes from the point of view of a science that is concerned with production relations and not with productive forces.

Specific characteristics in the development of productive forces, especially its pace, produce a specificity in production relations as well, because productive forces and production relations are inseparable and cannot exist without each other. This reciprocal tie between productive forces and production relations is the key to why a common mode of production is not able to eliminate differences in production relations completely, although it minimizes them. Therefore, even under an identical mode of production, there are differences between countries and regions in production relations as well.

Underestimation of the geographic environment results in a failure to understand many differences in production relations, since, in an indirect manner, the geographic environment influences production relations, especially the geographic division of labor and production ties.

By accelerating or decelerating the rate of development of productive forces, the geographic environment also accelerates or decelerates the entire pulse of societal life in a country or region. For this reason underestimation of the geographic environment also has a detrimental effect on the economic sciences.

An incorrect definition of the subject matter of economic geography leads to an erroneous orientation in economic-geographical research. Another consequence is that economic geographers, who, by the nature of their vocation, should struggle to maximize attention to local characteristics in geographic conditions in economic development and should equip people in applied work with knowledge of specific geographic characteristics, in certain cases come out with discourses on the general laws of development of capitalism and socialism; that is, they stray into another field.

To become convinced of this, it suffices to acquaint oneself, for example, with Ya. G. Feigin's book, *The Location of Production under Capitalism and Socialism* [*Razmeshcheniye proizvodstva pri kapitalizme i sotsializme,* Moscow, 1958]. Judging by the theme and considering that the author regards the location of production as the subject matter of economic geography, this work should provide a theoretical basis of economic geography as a science. But it is not difficult to see that this book belongs to the field of economics. It contains no specific regularities or even general propositions of geography. It points out no patterns of interaction between society and nature or territorial differences in these interactions. After reading this work, one could say either that it has no relation to economic geography or that our whole country is a uniform entity, with nature completely subordinated to decrees and resolutions that are capable of changing it in any direction without reckoning with its objective laws. One could name other essentially economic works that nevertheless claim to be economic-geographical. Yet the authors of these works do not consider economic geography a geographical science and include it in the system of economic sciences. For example, V. M. Volpe and V. S. Klupt write: "Economic geography is one of the economic sciences."[38]

A similarly incorrect view of economic geography is still held by some Soviet economists. In other words, one still finds in Soviet economic and economic-geographical literature statements denying the fundamental difference between economic geography and economics. Some scholars in our country still fail to see the difference between works in economic geography and works in the economics of countries and regions.

Thus, geography, and, above all, economic geography, is concerned not with production relations but with productive forces in their territorial expression and in combination with the instruments of labor. At the same time, economic geographers study their subject matter in connection with production relations, taking account of the influence of the latter (the mode of production) on productive forces. Geography, therefore, is an integrated science of the material foundations (natural and social) of the development of society, which is inconceivable without the geographic environment and its complex of natural and social conditions. Economic geography is concerned with the social conditions in which the process of social development takes place. But it does not specifically study production relations, and there are no grounds whatsoever to include it in the system of economic sciences.

Among the factors determining the development of the geographic environment, the most important are: the mode of production, the character of productive forces, specificity in territorial combinations of productive forces, and peculiarities in the effect of the laws of nature.

The requirement of a rigorously separate study of human society and nature leads to the notion that comprehensive cognition of the natural and social (economic) conditions around us is impossible. Criticism of a unified geography lacks, more often than not, a clear-cut definition of precisely how this term is understood. As a result, the broad approach in geographic research that includes territorial complexes developing under the influence of laws (natural and social) of different qualities is termed unscientific and is alleged to be geographical determinism.

" 'Unified geography' signifies a mechanical confusion of patterns of social development with the laws of nature. Because of its unscientific character, this confusion of patterns leads inevitably to a dead-end and retards the development of both physical and economic geography."[39]

Adherents of such views, who deny geography as a science, assume the possibility that either there are an infinitude of individual geographies without a common object of study, which are concerned with the arrangement on the earth of an infinity of things (soils, vegetation, textile factories, wheat crops, types of animals, and so on) or there are two geographies — physical and economic — which are totally autonomous (but related) sciences that are not connected directly with each other. The unification of economic geography and physical geography is declared "seditious," and the possibility of producing general geographical works is thereby totally denied. The idea that there are sciences that concurrently study natural and social categories and study combinations and interactions between nature and society is flatly rejected by some economists, and not only by economists. "How can one unite a natural science with a social science into a 'unified' geography? What kind of science is it that turns up between the social and natural sciences? Unification of these sciences leads to confusion between the laws of nature and social laws (otherwise the unification of these sciences is purely nominal), something that was opposed more than once by the founders of Marxism-Leninism and that is widely used by the bourgeois pseudoscholarly geographers — the Malthusians, geopoliticians and others."[40] This is the usual tone and spirit of the denials of the unity of geography, which are obviously devoid of even slightly scientific arguments. Unfortunately, such statements still occur today.[41]

In essence, the opposition to the unity of geography, regardless of the wishes of its exponents, is based on a rejection of the monistic view of the environment and on a denial of the existence of a causal tie between all the categories of the material world.

Such criticism of the notion of the unity of geography is based also on the raising to an absolute of the specific character of the laws of social development and therefore on the inevitable separation of human society from the rest of nature, on the dualistic contraposition of society to the rest of nature. Hence it follows, as a fully logical conclusion, that an insurmountable barrier runs between the social sciences and the natural sciences. By placing human society and the rest of nature in absolute opposition, the adherents of such views separate into two parts the material world of nature, which is actually unitary, and along with it geography. The adherents of such views, whom we call adherents of a split geography, in effect, deny determinism in the relations between society and nature.

This kind of artificial separation between the individual geographical sciences, above all between physical and economic geography, reflects a separation of natural (material) sciences from social (spiritual) sciences. The notion that cognition of the geographic environment as a whole is impossible is based on a dualistic incomprehension of the unity of the general and the particular, since the effect of the laws determining the development of the particular is raised to an absolute.

The split between the natural and the social in philosophy has been carried in certain instances to the point of efforts to establish a fundamental difference between the laws of natural science and the laws of dialectics, which, it is true, ran up against rather solid arguments from Soviet philosophers.[42]

The separation of human society from nature was viewed by Engels as "... a foolish and unnatural notion of some opposition between spirit and matter, man and nature, soul and body, a notion that spread through Europe during the decline of classical antiquity and reached its maximum development in Christianity."[43]

One of the justifications of the split between physical and

economic geography is the sharp difference in the definition of their subject matter. Whereas physical geographers, despite specific disagreements, unanimously declare a certain part of the material world to be the subject matter of their science, among economic geographers the notion is still widespread that economic geography is a science with location as its subject matter.[44]

It is perfectly clear that without location the process of cognition is altogether impossible. But the definition of a science's subject matter as the location of anything at all on the surface of the earth deprives that science of material content. The definition of economic geography as a science of "location" seems deeply erroneous to us.

Here are several examples of such definitions, in our view incorrect, of economic geography.

"Economic geography is a socioeconomic science that is intended to study the developmental patterns of a sphere of social phenomena such as the location of production in the conditions of various socioeconomic systems."[45]

"The economic geography of the Soviet Union is a new scientific discipline that is concerned with the patterns of the location of production, and the development and formation of economic regions under socialism."[46]

"We are among those Soviet economic geographers who hold that economic geography is a science of the location of production, where production is understood to be the totality of productive forces and their concomitant production relations."[47]

Contrary to all these definitions, we believe that the location of an object on a territory, be it textile factories or public production as a whole, cannot be understood without studying the object itself. *The location of material things can be cognized only by those sciences that study these things themselves.* For example, the economics of the textile industry cannot avoid studying its location, and so forth.

There does not and cannot exist a specific science that is concerned only with the location of a branch or a combination of branches of the national economy. The same is true for natural science. When studying soils, soil scientists cannot fail to study their location at the same time, and there is not and cannot be a specialized science of the location of soils.

The denial of the unity of geography and the separation and contraposition of its two branches has been so widespread that it was even set down in the decision of the Second Congress of the USSR Geographical Society, in February, 1955, which referred to economic geography as a specific science of the location of production. Here is a brief excerpt from that decision: "A geographical science that has acquired great importance is economic geography, a social science concerned with the geographic location of production (understood as a unity of productive forces and production relations) and the conditions and peculiarities of its development in various countries and regions."[48]

The Second Congress of the USSR Geographical Society did a great deal of useful work for the further development of the subfields of physical geography, but as far as theoretical propositions in geography as a whole are concerned, it was not productive. The Congress had extremely few reports on broad geographical themes that would promote the unification of geographers, and, conversely, too many reports on narrow, specialized themes that diverted attention from the basic problems of geography.[49] It was apparently this orientation in the work of the Congress that gave rise to its incorrect definition of the subject matter, and hence the essence, of economic geography.

The location of production is first and foremost a perpetual *process* that is inevitable under *any* mode of production. This is a territorial (in a broader sense, spatial) form of development of production. Without this form it is impossible for anything at all to exist and develop, since nonspatial matter does not exist, just as its nonspatial development is nonexistent.

The location of production is at the same time a highly important practical problem that is continually solved and continually reappears before human society.

In our country this is a highly important state problem to

which the Communist Party and the Soviet government have attached and attach enormous importance.

In contrast to the unplanned character of the process of location of production in the countries of the capitalist world, objective conditions are created in socialist countries for the planned, scientific guidance of this process in the interests of human society. The specific questions of the location of production in socialist countries should be resolved (and are being resolved) by special state agencies, primarily those in charge of the planning of the future development of the national economy (long-range planning).

The study of the location of production and the patterns of this location is indispensable, but this is by no means the job of only one science. It is impossible to resolve questions of location successfully by using the data of a single science. It is necessary to utilize and synthesize data from many sciences — economic, technical, geographical, geological, historical, and so forth.

Everyday life rejects the claims of some economic geographers to a monopolistic right to study the location of production as a specific subject matter, a specific science. In reality there is no such subject matter and no such science.[50] Indeed, how can one study production, say, from the standpoint of the technical or economic sciences without dealing simultaneously with its location? Can it really be that the economic and technical sciences are concerned with the economy of a country, and economic geography with the location of economy? That geology is concerned with minerals, and economic geography with their location; that economists study manpower, and economic geography its location? That the technical and economic sciences study transport, while economic geography studies the location of transport?

Acutally, the "locational" definition of economic geography has long since been disproved by life, since there is not and cannot be even a single concrete inquiry that is devoted only to the location of certain forms of production (just as they cannot be studied without reference to location).

Economic geographers, in fact, either study sectors of the national economy, often turning into specialized economists (who for reasons unknown call themselves economic geographers), or determine and study economic-geographic regions as territories with evolved complexes of economic conditions for the further development of production, that is, they study the economic complex of the geographic *environment,* disproving in practice the locational definition by producing specific, often highly valuable, geographical works.

It should be mentioned in passing that physical geographers who seek to move completely into natural science and the most active exponents of the locational conception of economic geography frequently arrive at a common view of geography, the conception of a split geography. This fact was pointed out a number of times by N. N.Baranskiy, who came out with great fervor against the distortions in Soviet economic geography.[51]

Those comrades who think the views N. N. Baranskiy criticized so sharply were done away with long ago are deeply mistaken. On the contrary, in a number of cases one can observe an increase in the tendency toward a split between physical and economic geography.[52] There is no question in our mind that physical geography continues to develop today in distinct isolation from economic geography, and the incorrect locational definition of the subject matter of economic geography is contributing to this in no small degree.

The monistic tendency in the development of geography finds expression in general geographical characterizations of individual countries and regions, and also in the development of integrated geographical research of an applied-science character.

The incorrect definition of economic geography as a science of location has roots going back to the indeterminist propositions in Kant's philosophy, and in even greater measure, to the Neo-Kantian concepts of members of the Baden school (Windelband and Rickert). The works of V. E. Den, who once created a school, were of great importance in the dissemination of the locational definition of geography. Den's views were very remote from Marxism and, to a considerable extent, were based on the idealistic aspects of Kant's philosophy. Den interpreted economic geography as a science without a definite object of study, as a supplement to economic history and as a specific supplement to political economy. He spoke of "... economic geography, which has the purpose of studying present-day phenomena in economic life in their geographic diffusion and this study can be carried out either by economic branches (agriculture, the mining, metallurgical and processing industries, trade, transportation routes, means of contact), or by localities (i.e., by countries or by parts or regions of a country). Economic geography is a kind of supplement to economic history: the latter studies economic life in the past, and the former studies it in the present."[53]

The theory of V. E. Den and his pupils did not differ in any fundamental way from the concepts that dominated bourgeois economic geography in Czarist Russia. "Bourgeois Russian economic geography, separated from physical geography by the ideological barriers of Neo-Kantian dualistic theories, was unable to cope with the new tasks of life posed by the socialist revolution."[54]

Den's definitions then became all but a basis for the exclusively economic definitions of economic geography, which, as we have pointed out, still survive today. These definitions by Den also presaged the formulations that became especially widespread in Soviet economic geography in the 1930s.

In studying the history of Soviet economic geography, O. A. Konstantinov gave what, in our view, is a totally correct description of the basic and general features of the school in Soviet economic geography that committed the sin of simplistic sociologizing: "... (a) idealistic separation of the location of production from the geographic environment, which is one of the constant and necessary conditions of the material life of society; (b) the almost complete absence of works on concrete themes, sociologizing and methodologizing on 'general' questions without relying on knowledge of factual material, incompetent and unfair criticism of other works; (c) the treatment of specific economic-geographic questions showed an inability to discern local specificity and a tendency to slide into general discussions."[55]

The sociologizing of economic geography in the 1930s did unquestionable harm to the discipline. The class and Party orientation of science were interpreted by certain economic geographers as a need for replacing economic geography with politics. It was no accident that in 1934 the Party and the government passed a special decision, "On the Teaching of Geography in Elementary and Secondary Schools in the USSR," aimed at defending geography against its sociologizers. This resolution retains its importance today, since all of its provisions have by no means been carried out. It was no accident that the twentieth anniversary of that resolution was marked by Moscow geographers with the publication of a special collection.[56]

It should also be recalled that even after the above-mentioned resolution was published, economic geography continued to be excessively sociologized, since some economic geographers believed that the resolution on the teaching of geography had a bearing on school but did not extend to the science as a whole. So on September 10, 1937, *Pravda* published an editorial sternly condemning the sociologizing school in economic geography.

The separation of the location of production from the geographic environment led inevitably to a denial of the objective character of the interaction between society and nature. The separation of location from specific forms of matter inevitably leads to geographical indeterminism, since it is invariably accompanied by an underrating of natural conditions that sometimes reaches the point of profoundly erroneous conclusions — the denial of nature as an objectively existing category.

In opposing the politicization of economic geography, I. A. Vitver wrote quite validly: ". . . The task of the Marxist geographer is not at all to replace geography with a summary of Lenin's works on imperialism and the program of the Comintern, but to present a geography that is illuminated by Marxist-Leninist theory and is politically oriented, but not replaced by politics."[57]

In summarizing this chapter, the following conclusions must be drawn.

1. The unity of geography is defined not only by a common object of study but also by a certain *unity in the method applied*. All of the geographical sciences, like geography as a

whole, investigate their subject matter through its *territorial complexes*. The geographical sciences can therefore be called *sciences of territorial complexes. The combination of the geographical method with specific subject matters,* all of which are either elements or complexes of elements of the earth's landscape envelope, *consolidates* the geographical sciences into a single *system,* thereby distinguishing it from the other scientific systems related to geography (economic, technical, biological, etc.).

Herein lies, in our view, the *methodological essence of the unity of geography.*

A specific form of the geographical method is *regionalization,* without which geographical inquiry is inconceivable. Another specific form of the geographic method is *mapping,* which is also an indispensable attribute of any specific geographical inquiry.

2. In studying the social (economic) elements of the geographic environment and their territorial complexes, economic geography evinces a significant (substantive) difference from the economic sciences, which study production relations. For this reason economic geography cannot possibly be categorized as an economic science, which would remove it from the system of geographical sciences. A classification of sciences should be based on a *combination of the subject matter studied by the given science* and the *basic method* it employs.

3. An incorrect definition of the subject matter of economic geography (as location) not only leads to the separation of economic geography from the geographical sciences, but also deprives it of a *material subject matter.* In essence, the locational definition of economic geography is a revival in somewhat altered form of the old chorological concepts that put chorology, that is, specificity in the approach to a study, in the place of *subject matter.* In other words, the material subject matter is replaced by methodological category.

The locational conception of economic geography develops into a rejection of the determinist view of the world. By raising the laws of social development to an absolute, it removes human society from the material world of nature. Therefore, the development of the locational definition of economic geography leads ultimately and inevitably to indeterminist conclusions.

CHAPTER SEVEN

Concerning the Boundary Between Geographical
Analysis and Geographical Synthesis.
The Scale of Inquiry in Geography.
Regional Geography as a Part of Geography.
The Problem of the General and the
Particular in Geography. Theoretical and
Practical Questions of Geography.

The scarcity of theoretical proofs of the definite unity of geography is one of the most important reasons for its lag behind
the ever-increasing demands of practice. Research limited to a
study of individual elements of the geographic environment
can be done only within the realm of the homogeneous laws of
its development. It is possible that this fact was the reason
most geographers have directed their efforts toward more intensive study of individual elements of the geographic environment, in which geography has achieved its greatest successes, since the lack of a general theory of geography is not
such a strong hindrance here to the process of cognition. But
even the specialized geographical sciences, particularly if one
thinks of their long-range development, are in need of broad
generalizations of their analytical research. But synthetic
generalizations of this kind, even in the specialized geographical sciences, can only be done blindly if there is no general
theory on the whole.

A considerably smaller segment of geographers has focused
its efforts on an integrated study of the individual parts of the
earth's surface, and there are even fewer geographers studying the geographic environment as a whole. This type of integrated geographical research is done only by representatives
of regional physical geography (primarily landscape scientists)
and also by regionally oriented economic geographers. One
should not be surprised by this. The lag of a theory of geog-

raphy has an especially strong impact on integrated works. Integrated research on the geographic environment has turned out to be a relatively backward area in Soviet geography. Yet it is precisely such research that is especially needed today in practice.

The process of differentiation of science has raised very pointedly the question of the relationship between the subfields of geography and geography as a whole. Here, once again, we encounter the problem of the relationship between analysis and synthesis.

It is common knowledge that by analysis of individual components of the earth's landscape envelope, singled out for independent study, the individual subfields of geography (geomorphology, climatology, agricultural geography, etc.) achieve a more profound understanding of these components. Analysis, therefore, is a necessary step in the cognition of the whole.

But where is the limit of analysis in geography? Does such a limit exist and if so, what part should it take in concrete geographical research? These questions are by no means simple ones, and their resolution is of great importance for research.

We contend that though analysis in general is boundless, there are bounds for geographical analysis. These bounds are determined by the object of geographic study itself. As long as individual elements of the earth's landscape envelope are investigated analytically as parts of a whole, the inquiry remains within the bounds of geography. But when an intensification of analysis leads to a loss of understanding of the subject matter (relief, climate, agriculture, etc.) as an element of the earth's landscape envelope, when analytical investigation of a *part* of the whole changes into an investigation of the *whole,* then the researcher is straying from geography into other areas of science (geology, physics, biology, economics, and so forth). In other words, the researcher encounters here a *qualitative* transition of one form of motion of matter to another, and then he will be dealing not with the patterns of development and interconnections of the earth's landscape envelope but with patterns and interconnections of a completely different character — those which determine the internal causes of the development of the individual elements of nature and society. We have already discussed the consequences this leads to.

The limit of geographical analysis is relative in character, and it cannot be understood in simplified terms, as some limit of cognition. It delimits not the process of cognition but the object of cognition. Therefore, the process of cognition is boundless even in geographical analysis.

The problem of the general and particular in geography is the problem of the relationship between analysis and synthesis. This general proposition clearly applies not only to geography, but here it probably finds one of its most vivid manifestations. Hegel has already pointed out, quite rightly, that division if "only *one* side, the unification of what has been divided is the most important."[1]

The problem of the irreducibility of the whole to the sum of its parts finds its solution in scientific synthesis. Based on the results of geographical analysis of the individual elements and complexes of elements of the earth's landscape envelope, geography is able, by means of synthesis, to create integrated images of the landscape envelope establishing the general regularities of its development, and also to study the interactions within it.

The data of many sciences are usually synthesized in the practical resolution of individual tasks. In the designing and construction of any large structure, the data of not one but at least several sciences are synthesized and utilized. The same is true in the construction of space rockets, hydroelectric stations, roads, and so forth. The possibility of synthesizing data from various sciences is a highly important factor in the development of technology and of all production.

Engineers, doctors, foresters, and agronomists make use in their practical activity of the achievements of not one but many sciences, synthesizing these data and thereby creating so-called applied sciences. With the development of the process of production, this kind of synthesis of the results of inquiry by various sciences, applied to the resolution of certain practical tasks, continually increases and expands, now often encompassing complexes of sciences concerned with completely different forms of matter and fundamentally different laws of the development of the material world. Thus, in the applied sciences, physics and chemistry combine with biology, biology with economics, physics with geology, and so on.

The conditionality and nonlinearity of the boundaries between sciences are thus manifested quite vividly in the application of science to the resolution of practical tasks. But the conditional character of the boundaries (or, rather, the intersections) between sciences does not mean they are nonexistent.

Therefore, synthesis *per se* should be distinguished from geographical synthesis, which is specifically limited by subject matter. Geographers only synthesize data on the earth's landscape envelope as investigated in terms of territorial complexes. But the results of geographical synthesis, in turn, can be synthesized together with the results of research in other areas of human knowledge. In practice, the results of geographical synthesis are primarily characterizations of territorial complexes of the geographic environment, that is, combinations of natural and social conditions that evolved on a given territory at a given time.

The differentiation of science without sufficient synthesis leads to the erroneous contentions that the demise of broad sciences is inevitable as a result of the development within them of specialized branches, which we discussed in previous chapters. For example, D. V. Nalivkin wrote: "... It would be interesting to know what classification of sciences will exist fifty years from now. One thing can be said: The sciences of geology and geography will cease to exist."[2] Such predictions will not come true. It would be possible if the whole were composed of the simple sum of its parts. However we know that the whole is *never* the simple sum of its parts, but is *always* a special *quality,* which is impossible to understand by means of fractionalization alone. For this reason the broad sciences, including geology and geography, will continue to exist fifty, a hundred, and even a thousand years from now, contrary to D. V. Nalivkin's prediction.

However geographers today are unquestionably confronted by the task of overcoming the lag in the area of geographical synthesis. Geographers must remember that "thought consists just as much in decomposing the objects of consciousness into their elements as in combining related elements into a unity."[3] There is a need for more boldness in the generalization of the accumulated results of analytic research.

An important place among synthetic geographical works is held by integrated geographical descriptions and characterizations of individual countries and regions. But it must be pointed out immediately that works of this kind usually do not extend beyond the bounds of geographical synthesis and constitute characterizations of the geographic environment of individual territories. Consequently, regional studies do not generate any distinct new science.

Still, it would be wrong to consider general geographical and regional-geographical studies completely equal. Although regional geography does not have its own distinct subject matter or its own distinct method of study, and therefore is not a distinct, autonomous science, it still has some distinctive characteristics. Regional geography is concerned not with the earth's landscape envelope, as a whole, but only with the geographic environment of individual countries and regions. It is therefore different from general geography. Moreover, regional geography is concerned with the geographic environment delineated by historical-geographical boundaries, which are most often state or administrative boundaries. This distinguishes regional geography from physical and economic geography, whose regional divisions are concerned with territories outlined by natural (physical geography) and social or economic (economic geography) boundaries. Like geography, as a whole, regional geography synthesizes the results of the investigation of all elements of the geographic environment, but only with regard to the specific territories of countries and regions. Geography, on the other hand, can synthesize data from the geographical analysis not only of individual countries but of the earth's landscape envelope as a whole. At the same time, geography can study and create synthetic generalizations of territories that are incomparably smaller than the territories of countries and regions, for example, landscape units, collective farms, enterprises, and so forth.

The earth's landscape envelope is studied by geography on various scales; in fact, the whole character of geographical study and of geographical works is highly dependent on the *scale of inquiry*. The scale of inquiry is of very great significance in geography. It determines in large measure the character and methodology of individual geographical works.[4]

As a general rule, the results of a study of one and the same object of inquiry carried out on different scales are *qualitatively dissimilar*. Therefore, geography actually is divisable not only into subfields (specialized geographical sciences) but also into parts, depending on the scale of inquiry.[5] Furthermore, there is a fundamental difference between the division of geography into subfields and its division into parts.

We have already discussed the nature of the division of geography into subfields. It leads inescapably to the emergence of relatively autonomous geographical sciences with their own subject matter and with their own methodological peculiarities. Each of these sciences is therefore capable of relatively autonomous development and establishes direct ties with associated, nongeographical sciences.

In addition to the specialized division in each geographical subfield, as in geography as a whole, there is a regional division, which does not generate any new geographical sciences. Depending on the scale of inquiry, such regional division generates not subfields but parts of geography.

The character of physical-geographical studies will not be affected by whether these studies are focused on a single landscape unit or encompass the entire earth; they will remain physical-geographical in character. The same may be said with respect to economic-geographical studies. Here differences such as those between geomorphology and climatology cannot arise.

If one speaks of the essence of regional geography and wishes to define its place in geography, one should be clearly cognizant of the fact that it is not a subfield of geography and not something indefinite that goes beyond its bounds, but *one of the basic constituent parts of geography.* In principle, regional geography is in no way different from general geography. Regional geography, like general geography, is concerned with territorial combinations (complexes) of the earth's landscape envelope in their development and interaction; *this is the geography of countries and regions,* one of the specific forms of geographical synthesis.

The place of regional geography, in geography as a whole, can be defined in a rather clear-cut way. Geography, as a whole, depending on the scale of inquiry, can be subdivided into three basic parts, each of which encompasses *all* of the branches of geography, that is, the entire system of geographical sciences.

Research that embraces the whole landscape envelope of the earth should be distinguished as a distinct part of geography — *earth science,* which is concerned with territorial differences in the scale of our planet and distinguishes continents and major orographic areas on its surface. Earth science should also concern itself with features in the world division of labor, in the geography of world population, in territorial combinations of world markets, the major world transportation arteries, and so forth. Cartography furnishes earth science with world maps and atlases — both general and specialized ones, but, as a general rule, drawn on a small scale.

Our conception of earth science, therefore, is somewhat different from the more widespread, predominantly physical-geographical conception, which we consider incomplete. Indeed, scholars in our country include only problems of physical geography under earth science. But the earth's landscape envelope can and must be studied also from the standpoint of economic geography and of geography as a whole. That is why we think it would be more accurate, for example, to give S. V. Kalesnik's widely known work, "Fundamentals of General Earth Science" [*O snovy obshchevo zemlevedenia*], a different title: "Fundamentals of Physical Earth Science."[6] Without question, general earth science should cover a much broader range of questions than has been the custom until now. It should include questions of the population of the globe, it should show the principal world centers of farming and livestock-raising, the major industrial concentrations of the world, and so forth.

Unfortunately, we do not know of a single work that with full justification could be titled "Fundamentals of General Earth Science." This situation can be explained, but it should not be justified. The absence of works in economic and general earth science is indisputably an indication of the inadequate development of geography, all of its achievements notwithstanding.

In studying the geographic environment of countries and regions, regional geography encounters their extraordinary diversity: in the size of territories, in the character of their landscape envelope, and in the sharpness of their internal differences. Therefore, within country-states there will often be a second taxonomic unit — a country (*strana*) that is a component of the state.

Within the second taxonomic unit — the country — regions outlined by historical-geographical boundaries are distinguished and studied.[7] Most often the boundaries of geographic countries and regions coincide with the boundaries of former precapitalist states or are determined by the settling of a certain nationality. But historical-geographical boundaries can also take shape inside a country with a single nationality if its individual parts were tenuously connected with one another for a long time because they had belonged to different states.

Furthermore, regional geographers distinguish and study regions outlined by historical-geographical boundaries only if these regions retain a geographic specificity at the time of the study.

Examples of such regions are Normandy and Brittany in France, Bavaria in Germany, Scotland in Great Britain, the Tatar Autonomous Republic in the Russian Federation, and so forth. These and similar regions are not economic regions (although such congruence may occur in the form of exceptions), nor are they physical-geographic regions. But they objectively exist and have many specific characteristics in nature, population, economy and cannot be omitted from specialized geographical study.

These territorial units are what we adopt most frequently as the lowest taxonomic unit of regional-geographical study, distinguishing at the same time the basic differences within them, but without necessarily carrying out additional, more detailed regionalization.

The boundaries of regions studied by regional geography often coincide with the boundaries of territories on which the process of nation formation took place. *"... A nation is a historically established, stable community of people that arose on the basis of a common language, territory [our italics —V.A.,] economic life, and psychological mold that manifests itself in a common culture."*[8] *Territory*, that is, the geographic environment in which the nation formed, unquestionably influenced the specific characteristics of its respective nation.[9] The connection between geographic conditions and nation formation is completely obvious to us. A nation can take shape only after a long period of regular social intercourse, only as a result of corporate life over a number of generations. This long-term corporate life inevitably implies a certain territory, with certain established geographic (natural and social) conditions, which unquestionably leave their imprint *through production* on the nation that is forming and on the peculiarities of it that we sometimes call *national character*. The nation, in turn, while forming on a certain territory, also influences the direction of development of the geographic environment *through production*.

It is therefore not surprising that, as a result of this interaction, national geographic boundaries define regions with specific characteristics in the geographic environment.

Regional-geographical work cannot be done with territories that are too small and do not represent a customary, historically evolved unity. Objects of regional-geographical study cannot be defined arbitrarily, they actually exist; and it is the task of geographers to discover them for future study.

The boundaries of national-state entities are, in our opinion, the basic regional-geographic borders. But since elements of the nation, specifically the territorial and linguistic-cultural communities, existed prior to capitalism, a historical-geographical division of the earth's surface also existed before capitalism. Regional geographers therefore deal with countries and regions whose boundaries have different origins, although in every instance they are historical-geographical. Today historical-geographical boundaries can demarcate pre-capitalist, capitalist, and socialist countries and regions.

Thus, we define country and region, in general geographical terms, as territories distinguished by historically evolved geographic characteristics. They have a definite historically established unity of nature, population and economy, a unity that has retained its significance up to the time the given country or region is studied. The historical unity of individual territories in our country is usually taken as the basis of the demarcation of administrative boundaries. The boundaries of provinces, territories, and particularly republics in our state are far from arbitrary; they define only certain regions with a natural combination of characteristics in customs, a distinctiveness in national composition, and some minimum economic integration. The same criteria may be observed in other countries.

It is very interesting in this respect to compare the present administrative-territorial division of a country with the feudal-state units that existed on the same territory. It turns out that in most but not all cases the boundaries of feudal states that long ago became extinct retain their significance for a considerable period of time, remaining historical-geographical boundaries. Even in such countries as France and the USSR, where the foundations of feudal society were shattered most radically and where administrative divisions underwent the most substantial changes, we can still see some link between the present administrative divisions and the fiefs that formerly existed. The departments of bourgeois France seldom cut through the boundaries of historical provinces. They are usually grouped together on the basis of those provinces: Normandy, Burgundy, and others. The borders of the central provinces of socialist Russia are frequently quite close to those of the old Russian principalities. This phenomenon is not an accident and is not a harmful anachronism. It attests to the viability of state boundaries, which retain their historical-geographical (and therefore, to some extent, economic) significance long after the states they defined have faded into the past.

Thus, in contrast to physical geography, which uncovers physical-geographic regions, and in contrast to economic-geography, which reveals economic regions, regional geography is concerned with country-states, countries (even those that are not states), and historical-geographic regions, that is, it is concerned with the geographic environment of territories defined by state and historical-geographical boundaries.

Although any geographical inquiry is unthinkable without the use of the geographical method, in regional geography the geographical method probably finds its fullest expression. Without the geographical method, one can obtain only a set of sundry, totally unrelated bits of information. No regional-geographical characterization, not even the most primitive regional-geographical description, is possible without the geographical method.

N. N. Baranskiy, who was one of the first Soviet geographers to affirm the necessity of regional-geographical works, wrote in this connection: "The thrust of our proposal is to create by no means mechanical amalgams but *characterizations* that contain a logical combination of each country's most important distinguishing features and to fit their features together as closely as possible."[10]

Although we are in full accord with this statement by N. N. Baranskiy, we cannot possibly agree with him when he says that regional geography "... should be merely an organizational form of combining sundry knowledge about a certain country."[11]

Regional geography is not a distinct science, but it is not an organizational form, either; it is the same general geography that N. N. Baranskiy mentions in the article just cited. Regional geography does not replace physical and economic geography, but combines their data for synthetic characterizations of countries and regions. This is one of the principal forms of geographical synthesis.

This conception of regional geography cannot possibly be considered narrow. Geography has every opportunity for a broad and comprehensive approach to the countries it studies. Information about nature, population, and economy, about peculiarities of historical development, political and cultural life, are combined in regional geographical works, affording in totality a more or less complete and integrated (synthetic) image of the country and of the differences within it, as well as pointing out and explaining the causes and character of the latter.

When he studies countries and regions, the geographer inevitably uses data from sciences other than geography. But all of these data are taken within a general framework, with the purpose of depicting as fully and comprehensively as possible the country's geography, that is, the territorial complexes of the geographic environment that have evolved there, and explaining the geographic features of the country or region. These data include information about the country's history, material describing traits in its culture, political life, and so on.

However all the data from other sciences are taken with the purpose in mind of showing and explaining the *geographic* characteristics that have evolved in the given country. All of the sundry material of regional-geographical work should be grouped around a central core, around the major elements that characterize the country, its geographic environment, and the territorial peculiarities in that environment. This is the basic *content* of regional-geographical works. It is impossible to compose regional-geographical works without describing nature, population, and economy in their unity and their territorial complexes. Other information is important, but secondarily so.

There is really nothing that would make it possible to speak of regional geography as something *outside* of geography as a whole, as some distinct organizational group of various sciences. Attempts at this kind of organizational unification of different sciences on an equal basis can yield nothing but a mechanical set of diverse information. At best it will be something encyclopedic, but not integrated characterizations of countries and regions. Such organizational unification may lead to a regional geography that N. N. Baranskiy quite rightly called a nightmare.[12]

Unfortunately, there is still a widespread, totally incorrect conception of regional geography as something indefinite and not formalized, yet at the same time as something broad, extending far beyond the bounds of geography. But as we have pointed out, regional geography does not have a distinct subject matter that is different from geography, it does not have its own distinct method or its own distinct purpose. One could not identify anything substantial that would distinguish it from geography in any fundamental way. The objects of study — specific countries or regions — are geographical objects of study. The method of study is the geographical method. The goal of study ultimately reduces to composing characterizations of the landscape envelope that took shape on the territory of countries and regions. In other words, there are no differences here, either, from the goals that geography sets for itself.

The use in regional geography of data from other sciences cannot possibly, of course, lead to the conclusion that regional geography is broader than geography as a whole. All of the data from other sciences are of interest to regional geographers only to the extent that they help to understand the geographic specificity that evolved in a certain country up to a certain time. Physics, for example, makes wide use for its purposes of mathematics, but does not cease to be physics because of this. Chemistry uses data from physics, but remains chemistry. Geography makes wide use of data from geology, economics, history, and other disciplines, but does not cease to be geography because of this.

The use of data from other sciences does not distinguish regional geography in any way and does not allow one to speak of it in any degree as some distinct science or as some embryo of a new science that has not yet taken form. The wide opportunities to make use of the achievements of one science by other sciences only proves once again that there are no impenetrable partitions between sciences and proves that science is the tree "... of living, fruitful, veritable, mighty, omnipotent, objective, absolute human cognition."[13]

In speaking about regional geography, one cannot fail to emphasize its practical significance and the necessity this engenders of giving it a theoretical basis. Without such a basis, without elaborating at least the most fundamental methodological questions, it is hardly possible on the present level of scientific knowledge to produce a "Grand Geography of Our Homeland" and a "Grand Geography of the World"; and the compilation of these works is at present a matter of state importance and a matter of honor for Soviet geographers.

The scornful attitude toward regional geography, the widespread opinion that a scientific regional geography is impossible because of the wall that supposedly stands between the natural and social sciences, the unsuccessful efforts to train regional specialists without a geographical education — all this hampers the study by Soviet geographers both of their own country and of foreign countries.[14] All this has produced a situation in which we now have few (and the number continues to diminish) geographers with a broad background who are able to work in the area of regional geography. In this way indeterminist distortions in theory have led to pernicious consequences in yet another area of geography — regional geography.

The solution of the problem of the general and the particular (with reference to geography) is of very great importance for the further development of regional geography. It is well known that the formation of the geographic environment and its individual elements is related to the general laws of development of nature and society. The formation of the social elements of the geographic environment is decisively influenced by the mode of production and the character of production relations. As we have repeatedly emphasized, the general laws of development of the geographic environment cannot lead to its complete uniformity over the entire surface of the earth. So if a geographer wishes to show the "face" of a country and write a regional work, he cannot possibly ignore local characteristics and specificity either in nature or in population and economy.

We know that all general features inevitably manifest themselves in specific ways. A general determining factor exists and operates in every specific instance not apart from particulars and manifests itself in specific characteristics, in the form of features that are typical of a given country or region. The geographer's task is to show the effect of the leading factor in its specific manifestation without "drowning" in facts.

In any geographical work it is extremely important to single out the principal and typical features that constitute the *basic content of geographic specificity,* which can vary quite considerably from country to country and region to region. When revealing geographic specificity, one inevitably reveals the indication of leading factors on the territory of the given country or region. When a regional work presents geographic specificity correctly, that means that it correctly depicts the effect of leading factors, the effect of general laws that determine the development of production.

Some comrades, as was mentioned earlier, do not wish to understand this proposition, which would seem to be a perfectly simple and clear one. For example, Ya. G. Feigin, who alleges without evidence that the unity of geography inevitably entails a mechanical confusion of the patterns of social development with the laws of nature, even organized a kind of poll by quotation on the question of whether the general is more important, or the particular.[15] He cited a series of quotations from works by the founders of Marxism that point out the importance of studying particulars, and a series of other quotations on the importance of studying the general.

If one considers the quotations cited by Ya. G. Feigin, one cannot fail to conclude that it is impossible to divorce the particular from the general or the general from the particular. Feigin himself does not draw such a conclusion, but he sets the general at variance with the particular, criticizing geographers who dare to show geographic specificity.

Criticism of showing specificity, which is supposedly based on a defense of the general laws of social development, cannot produce anything useful for geography, and especially not for applied work. It merely justifies theoretically the stereotyped approach in economic management and increases the tendency to ignore local characteristics in natural and social conditions in the planning of the national economy. This criticism, which is based on contraposing the general to the particular is attributable to the fact that its authors have completely forgotten Lenin's thesis that "... the general exists only in the specific, through the specific. Everything specific is (in one way or another) general. Everything general is (a particle or an aspect or an essence) of the specific .... Everything specific partly belongs to the general and so forth .... Everything specific is linked by thousands of transitions to other *kinds* of specific (things, phenomena, processes) ...."[16]

The fight against showing geographic specificity has done great harm to the development of the national economy, particularly the development of agriculture, where consideration of zonal and regional differences is especially necessary and where, consequently, the regional approach is extremely important both in planning and in administration. "By the very nature of farming, it is transformed into commodity production in a special way, which does not resemble the corresponding process in industry. The processing industry divides into separate, completely autonomous industries that are devoted exclusively to the production of one product or a part of a product. The farming industry, on the other hand, does not divide into completely separate branches, but only specializes in the production of one or another market product, whereas the other facets of agriculture adapt to this principal (i.e., market) product. Therefore the forms of commercial farming are distinguished by a gigantic diversity, changing not only from region to region, but also from farm to farm. Therefore, when one examines the question of the growth of commercial farming, one cannot possibly limit oneself to broad data about all farm production.''[17] Can one consider this thesis of Lenin's relevant only to the capitalist economy and ignore it in the practice of socialist construction?

Geographers show the interrelations between society and nature with concrete material. But this is possible only if he singles out the main features, that is, that he ascertains and shows geographic specificity, without losing sight of the totality of the whole. This is the application in geography of V. I. Lenin's theory of the chief component. Geographic specificity is the chief component in geographical works on countries and regions.

Remembering that the general reveals itself in the specific or particular, geographers seek to show the specificity, that is, the totality of the whole, as completely and thoroughly as possible and to show the geographic environment as it has formed on the earth's surface. If someone succeeds in this and at the same time correctly understands and presents geographic specificity, he will also depict correctly the effect of the general and basic laws that determine the development of production.

Without specific works presenting the local characteristics of countries and regions, there is no geography. *The continuation and development of specialized analytical research along with synthetic generalizations of the materials from this research and the subsequent compilation of regional works based on these generalizations are what constitute the principal trend in the work of Soviet geographers.* And if individual geographers, in following this path, commit and probably will continue to commit blunders and errors, their work will still contain materials, conclusions, and generalizations that are beneficial to science.

Marxist philosophy equipped science with a consummate method of scientific cognition, the method of dialectical materialism. For geography this opened up new prospects and unprecedented possibilities for development, which are still being utilized inadequately. Materialist dialectics makes it possible to approach geography's object of study correctly as a *unity of heterogeneity.*

We can now speak of the unity of the earth's landscape envelope without the inevitable mechanical carry-over of the laws of nature into the sphere of social relations or of the laws of social development into the sphere of nature. Marxism makes it possible to understand the essence of the qualitative differences between nature and society, which develop, on the one hand, under the influence of general laws, and on the other, under the influence of their own specific patterns.

A correct understanding of these specific patterns makes it possible to examine the natural and social elements of the landscape envelope in their interconnection and interdependence. Materialistic dialectics makes it possible to understand correctly the character of the reciprocal influences between nature and society.

Geography can now study its material object of inquiry, the earth's landscape envelope, without breaking it up mechanically into parts and components, but also without confusing them, making sure to distinguish the specificity in their development and in the laws governing this development.

Thus, by applying the laws of materialistic dialectics in geography, it is possible to substantiate theoretically *the unit of geography* as a science with its own distinctive character in the combination of a common object of study with the method of its investigation. Other sciences do not have such a combination of subject matter and method as does geography.

As we have tried to show in the foregoing discussion, geographers have usually either seen the community of the geographical object of study but failed to understand the nature of the difference within this community or have seen only the specific subject matters of the individual geographical sciences, but failed to see the entire diversity of ties between them, failed to understand the community between these specific subject matters and failed to conceive of them as parts of a whole. This philosophic problem of the whole and its parts has still not been solved in geography. But as Marxist-Leninist materialist dialectics is creatively applied in geographical research, the conditions for its solution are being created. The problem of the whole and its parts is of special importance for the broad sciences. Without solving this problem in geography, it is impossible to understand its subject matter and hence the unity of geography as a science.

It should be pointed out that even among geographers who do not accept geography as a unified science, one can find an understanding of the importance of the problem of the whole and its parts for geography, and basically correct assertions about the essence of this problem. For example, I. M. Zabelin is correct when he writes that "... the internal differentiation generated by the process of development by no means violates the integrity of the phenomenon; on the contrary, simplification of the structure, at least at high levels of development of matter, attests to the approaching breakup of the whole. Integrity, therefore, does not presuppose absolute qualitative homogeneity, and this is another nuance, which makes it possible to distinguish the category of 'quality' from the category of 'integrity.' "[18]

It is for this reason that we can legitimately consider in definite unity such fundamentally different forms of matter as nonliving, living, and social, which together constitute the whole of the material world. However, a whole that consists of forms of different qualities is not, of course, the sum of these forms, but always more than the sum, which is a result of the process of interaction within the whole between its elements.

We also agree completely with I. M. Zabelin that one cannot regard everything contained in the whole as its parts. Only those elements, of course, that reflect the organic ties and interactions within the whole and are internal causes of the development of the whole can be considered parts of the whole. Therefore relief is a part of the landscape envelope (or an element of the geographic environment), but the individual minerals are not elements of the geographic environment. Human society is a part of the landscape envelope, but individual persons are not parts of it. In excluding individual minerals, individual types of vegetation and animals, and other such minutiae from the sphere of interest of geography, Zabelin is essentially speaking about the boundary of geographical analysis; and he thereby cites one of the major arguments in favor of the unity of geography, contravening his own previous assertions, since it is precisely the boundary of geographical analysis that separates the geographical sciences from the nongeographical (his examples separate physical geography and its branches from mineralogy, botany, zoology, and so forth). We discussed this kind of boundary of geographical analysis earlier.

As far as we are concerned, there is no question that the earth's landscape envelope really does exist as a "unity of the heterogeneous," and only in that capacity is its cognition possible. The qualitative differences within the landscape envelope not only do not destroy its integrity, but on the contrary, increase this integrity. Hence the exclusion of man from the landscape envelope on the ground that human society is "a special type of integrity, subject to its own laws of development . . . ,"[19] is clearly at variance with the correct conception of integrity as a "unity of heterogeneity." I. M. Zabelin does not exclude the animal world from the earth's landscape envelope, and, after all, that is also a special type of integrity, subject to its own laws of development that are fundamentally different from the laws of development of relief or of the atmosphere. If one adopts the position of denying the possibility of integrity in every instance where the development of the elements of a whole is subject to its own specific laws, then it is inescapably also a denial of determinism in the material world of nature, and matter turns out to be not unified but fragmented. With this approach, the unity of geography is completely denied.

Of no lesser importance at present is the problem of introducing the results of geographical reasearch, geographical methods and ideas into economic construction.

It is common knowledge that practice is not only the basis but also the goal of all cognition. It is the criterion of truth. It completes the cycles of cognition, proving the objective truth of a theory, and at the same time serves as the basis for the creation of new theories.

However practice does more than test the validity of a theory.

The very purpose of a scientific theory, in essence, is determined by practical needs. "Practice — cognition — practice again — cognition again — this form is endless in its cyclical recurrence, but each time the content of the cycles of practice and cognition climbs to a higher level."[20] And one must not fail to note that the criterion of practice is a historical category, which is therefore relative in character. Practice is not capable of completely confirming or disproving cognition at any given moment. Although disproving it at one stage of historical development, practice can confirm it at another. Thus, no matter how important the criterion of practice is, it cannot be raised to the rank of an absolute criterion, and it must be approached in historical terms.

At various stages of the historical development of society, practice has shown that underrating the significance of the geographic environment in the sphere of social development, like underrating local characteristics in the geographic environment, is a most flagrant mistake.

Economic practice in our country has also shown a number of times that the study of the natural elements of the geographic environment without reference to its social elements leads likewise to gross errors.

One could cite many concrete facts confirming these propositions. We shall confine ourselves only to a few of the best-known. One of these examples is the nihilistic attitude toward consideration of the local characteristics of the geographic environment in the proliferation of grassland crop rotations virtually throughout our country.

"Some have tried to prove that the grassland system is a guarantee against all mishaps and assures high crop yields in every region, regardless of soil and climatic conditions.

"Instead of applying Williams's theory creatively, it has begun to be turned into a dogma, and there have been attempts to adapt it to the arid regions of the south and to extend it throughout the territory of our boundless Soviet Union. As was to be expected, these attempts proved untenable."[21]

The nationwide proliferation of grassland crop rotation without proper attention to differences in geographic conditions and after N. S. Khrushchev's criticism the cessation of its use in areas where it had fully justified itself; individual instances of the planning of hydroelectric stations without proper attention to the damage that the flooding of settled and developed territories would inflict; the felling of timber without consideration of the conditions for its regeneration; the introduction of sub-tropical agricultural crops in zones where they are not economically justified; fishing by methods that lead to total extermination, — all these occurrences, which do serious harm to our national economy, can continue in the future if the underestimation of the influence of the geographic environment and the geographic factor is not eliminated.

Taking account of changes in the geographic environment that occur during the creation of large-scale industrial enterprises and as a result of industrial and urban construction is of very great economic importance. For example, Yu. G. Simonov writes: "The modern-day large city, with its complex of industrial and civil structures, changes the natural conditions of its territory. Cities take on 'their own' relief and climate, new drainage conditions take form, and so forth; in a word, a new landscape emerges, often drastically different from nature's landscape. At the same time, the quantitative and qualitative relationships that existed previously between components of the landscape are inevitably ruptured on urban territories. This rupture usually takes place over a very short period, which is attributable to industrial construction methods."[22] Yet there is still inadequate attention given in construction to landscape changes of this kind, which lead in certain cases to the destruction of structures just completed and sometimes uncompleted, shorten the periods of operation of buildings and necessitate premature capital repairs on them.

Another important problem of the national economy whose solution is impossible without the participation of geographers is the qualitative appraisal of lands for practical needs. It is especially necessary to solve this problem at the beginning of construction projects and in agricultural production. The planning of large-scale construction requires data both on individual elements of the geographic environment and integrated data on the region as a whole where the construction is to take place. Moreover, it is essential in the choice of construction sites to take account not only of the immediate needs of construction and of the future operation of the completed enterprise but also of the needs that will arise after the enterprise is put into operation. For example, "... when locating enterprises on territories undergoing new development, construction sites are often chosen from among the best farmlands. This frequently occurs at construction projects in Siberia, especially in the taiga zone, where small fields lie among vast tracts of forest. They were won away from the taiga at the cost of almost superhuman efforts by the first settlers in Siberia. The soils of these fields have been adapted for crops and differ substantially and favorably from the adjoining forest soils, which are deficient in humus .... Free of timber, easy to scan, evened out by many years of tillage, they are quite susceptible to regional reconnaissance and probably for this reason are easy 'prey' for planners. It is these fields that are chosen first for industrial facilities. To justify these illogical activities, the thesis is advanced that the revenue obtained from agricultural use of these lands is much less than the revenue that will be obtained upon completion of the industrial construction. The untenability of such arguments is not immediately apparent, especially at the planning stage of these structures. But much later, when the industrial enterprises go into operation, a need appears for lands near growing cities on which to create a vegetable and dairy base for these cities. This necessitates rooting out more timber and redeveloping from scratch lands that are less suitable for agricultural production."[23]

Agricultural production incurs enormous injury from incorrect estimations of the conditions of the geographic environment. Unfortunately, it is sometimes still believed that knowledge of soil and climatic conditions or even only of soils and vegetation is enough for the successful management of agriculture. But regardless of the potentialities of soil science," ... many factors affecting tillage conditions and harvests can be evaluated and 'measured' directly much more fully and reliably. In addition to soils and, of course, soil-forming rocks, large-scale maps of collective farms, state farms, and so forth, should show relief; and the description should interpret its role in the redistribution of heat and moisture, in the development of erosion under a given manner of working the land, and should thoroughly elucidate the significance of relief as a factor that facilitates or impedes the operation of farm machines."[24] "Correlating the yields of cultured and wild plants both with the natural properties of the lands and with the system of their use is the only means of fully ascertaining the relative profitability of various crops on given lands and the only way to disseminate with confidence the experience of advanced farms among all the other farms with similar or kindred conditions."[25] Research in the appraisal of lands for their better use in agricultural production is one of the practical confirmations of the impossibility of studying the geographic environment while approaching it from the standpoint either of natural science alone or of social science alone.

The appraisal of lands is possible only with an integrated, general geographical approach to their study, only with a synthesis of data on the natural and social elements of the geographic environment. It is no accident that research in the appraisal of lands leads geographers to conclude that geography is unified. Whether they wish to or not, geographers engaging in the integrated study of lands for the purpose of making agricultural appraisals of them arrive at approximately the same views about geography as those defended in this book. In fact, one can probably agree as well that such work contributes to a revision of views about geography to no less a degree than specialized theoretical research, since concrete general geographical research demonstrates *in practice* the whole untenability of theories that divorce the natural subfields of geography from the social, and the harm produced by the infiltration of these theories into the ranks of the managers of economic projects.

Integrated geographical research is of very great importance in regions of planned hydroelectric construction, especially on lowland rivers; that is, construction involving the flooding of large areas. In a number of cases the plans for hydroelectric construction have been drawn up without proper evaluation of the effect it will have on the geographic environment of the surrounding territory. Hence there came the inevitable underestimation of the deleterious consequences that often resulted from the construction of hydroelectric stations and the underestimation of the actual cost of the hydroelectric stations' electric power, the cheapness of which proved to be highly nominal, since the calculations did not take into account the large, often irreparable losses that were suffered as a result of such construction by other sectors of the national economy, especially agriculture, forestry, and fisheries.[26]

Planning the development of our country's national economy cannot be successful without recognition of the unity of natural and social (economic) conditions. "The ideological groundwork [of economic geography — V. A.] was the GOELRO plan and the work of the State Planning Committee in the 1920s in economic regionalization, based on an integrated, monistic conception of the world and on Marxist philosophy."

"The problem of relations between the productive forces of labor, the material and technical base of production (on the basis of its electrification), and the natural environment of regions was resolved in the above-mentioned works on the level of philosophical and practical unity and the close interconnection between them, and with specific examples. Obviously, this indirectly prepared the way for a correct resolution, based on Marxism, of the question of relations between physical and economic geography, in the sense of the necessity of their *interconnection* and fundamental unity."[27]

For a number of years a nihilistic attitude toward the geographic environment was propagandized in our economic, and sometimes even in the geographical periodical press. Under the banner of a struggle against geographical determinism, the idea frequently ran through the press that, in general, our country, with its new socioeconomic structure and advanced science, does not need to take account of the geographic environment or local geographic specificity; and the appeals of individual scholars who proved the harmfulness of such an attitude toward the geographic environment were dismissed as a manifestation of simplistic geographism. Moreover, the writings of scholars who urged that local natural and economic conditions be reckoned with sometimes encountered destructive criticism.

For confirmation, we cite an example that vividly depicts the character of such criticism. In 1947, S. G. Kolesnev put forth in the press a totally valid demand for more consideration of local conditions in order to reduce the amount of socially necessary labor in agricultural production. He wrote that "... agricultural products should be produced primarily in places where the amount of socially necessary labor per unit of output is minimized, that is, where the cost of the production and transportation of a unit of output is minimized .... To produce as much output as possible in Soviet agriculture does not mean at all that it should be produced at any level of expenditures. On the contrary, it is necessary to produce more with the minimum outlays of socially necessary labor per unit of output. However, conditions may evolve with respect to certain types of output such that production at minimum cost will not be able to meet society's full need for that product. Under such conditions society will be forced to produce in worse conditions, with greater outlays. However, under these conditions as well, it will seek at the same time to reduce costs by perfecting technology and by making all kinds of improvements."[28] In the same year of 1947, V. S. Nemchinov, examining questions of the location of agricultural branches, wrote: "the question of the location of crops and livestock-raising sectors may be approached in two ways: (1) as a problem of agricultural natural-historical and agroclimatic regionalization (specifically, in terms of determining the effect of climate on individual production sectors); (2) as a planning problem of the rational location of agricultural production (specifically, in terms of establishing the criteria for rational location of agricultural production)."[29]

These quotations, which speak of the need for taking account of local characteristics in natural and economic conditions, were *especially timely* in those years, since it was precisely at that time that agriculture was suffering acutely from stereotyped management, which was not taking account of the specific character of the geographic environment in the various zones and regions of our homeland. Unfortunately, these completely valid writings not only failed to obtain support, they were even sharply criticized by some economic geographers. Here, for example, is how V. F. Vasyutin interpreted the passages we quoted from the works of Kolesnev and Nemchinov: "Thus, in his work, 'The Organization of Socialist Agricultural Enterprises' (1947), Kolesnev bases the location of agricultural production on the anti-Marxist theory of minimum costs . . . ."[30]

The same writer goes on to accuse Kolesnev of regarding the movement of farm crops into new regions as impractical, since that would increase agricultural production costs.

V. F. Vasyutin wrote that ". . . This 'theory' of Kolesnev's is based on that same simplistic geographism, overestimation of the natural geographic environment and underestimation of the role of Soviet and Michurinian agronomy and of new agricultural technology in overcoming unfavorable natural conditions . . . ."[31] ". . . Academician Nemchinov ascribes decisive importance here to the forces of nature . . ., placing himself in the ranks of the advocates of a simplistic-geographical methodology in location."[32]

Of course we can overcome unfavorable environmental conditions. But scientific experiments and economic practice must not be confused. If it were harmful to impede scientific experiments in the introduction, for example, of grapes in the central regions of the Russian Republic, it would be even more harmful to plan the location of vineyards in these regions before grapes became an economically profitable crop there. Of course, one can also grow citrus fruits in the tundra, but how much will that cost? And wouldn't it be better to plan the location of citrus fruits for the time being in the southern Caucasus?

The necessity of rigorously taking account of geographic conditions in practice is not only not diminishing, it is increasing even more.

The development of the socialist economy during the creation of communism requires a transition to more perfect forms of organization of the material and technical base for management and planning of the national economy, and the introduction by the Party and the government of industrial and construction management based on economic administrative regions was, in particular, one of the forms of transition to territorial management of the national economy.

The geographical method is being increasingly introduced into economic practice, especially since the development of all branches of the economy is becoming increasingly integrated.

The leaders of the Communist Party of the Soviet Union continually point to the necessity of taking account of local characteristics in natural and economic conditions when solving questions of economic management. For instance, N. S. Khrushchev recalled this during his conversation with Soviet and Indian specialists at the Bhilai Metallurgical Plant. He said: "As to whether your plant will expand substantially in the future, that depends on the possibility of providing it with good ore, dolomites, water, everything such a large metallurgical combine needs. If such possibilities exist here, then it will be profitable. Back in the years when I was a student at the Industrial Academy, Professor Baranskiy gave us a course in economic geography, in which he explained what data should be taken into consideration when choosing a place for the construction of an enterprise."[33]

The scope of construction and its integrated character, the task of maximum savings in social labor and gains in time, more energetic economic use of resources in the eastern regions — all this requires a scientifically based territorial integration of productive forces in the process of their future development and a correct solution to the problem of their geographic location.

Consequently, a science that is specifically concerned with the geographic environment and is thereby capable of rendering considerable assistance to practice acquires great currency. This need for geography is felt in all of the principal branches of the national economy, which cannot be viewed in isolation.

In the area of agriculture, for example, the farming and livestock-raising systems are becoming complexes of agronomical, zootechnical, and economic-organizational measures that maximize the economic profitability in the given concrete natural and economic conditions, that is, in the given geographic environment. "Specialization and location are designed to assure utilization of the local natural and economic conditions that are most favorable for the production of given agricultural products."[34]

For people engaged in the most diverse areas of the national economy, it is very important to have sufficiently detailed characterizations of the geographic environment, which has evolved and is evolving in various parts of our large country in somewhat different ways.

"Many major shortcomings in the development of our economy occur because the managers of certain sectors of the economy sometimes forget about one of the renewable resources of large-scale reproduction and forget about the need for more integration of all branches of the economy on a region-by-region basis, with full utilization of the local natural and economic conditions that are accessible at a given level of technological development. It is this oversight that leads to stereotyped management of the economy (especially in long-range planning) and leads to substantial wastes of labor, hampering the further growth of its productivity."[35]

In essence, the inadequate study, and hence the inadequate consideration in practice of geographic conditions means ignoring reality. In order to actually take account in practice of the whole diversity of local characteristics, which would unquestionably bring enormous additional advantages to our economy, it is imperative to do away with the scornful attitude toward geography, which is often viewed merely as an academic subject necessary for a general education. It would seem perfectly obvious that taking account of local characteristics and geographic specificity is impossible without a comprehensive study of these characteristics and this specificity. Neither specialists studying individual aspects of the economy (engineers, agronomists, economists) nor specialists studying individual components of nature (climatology, hydrologists, soil scientists, etc.), despite of all of their indisputable specialized erudition, are capable of providing characterizations of the geographic environment.

By relying on the results of specialized research, broad-based geographers should and can give integrated characterizations of natural and social conditions, on the condition that they take a regional approch to the territories they study.

The latter thesis is perfectly obvious. We repeat only that now that economic construction is usually carried out in the form of creating not so much individual industrial enterprises as production complexes that sometimes cover vast territories, sets of specific, disjointed geographical materials (on relief, climate, population, transport, etc.) are not able to satisfy the direct needs of practice. True, economic practitioners do not always understand this yet, believing sometimes that a modern economy can also be developed the old way, without knowledge of geographic complexes.

The further we proceed along the path of the construction of Communism, the more we must have comprehensive knowledge of the geographic conditions of our large and diversified country. Social influence on the geographic environment in a socialist society is a qualitively new process, since the scale and forms of this influence and the scope and pace of transformational work are completely different, incomparably greater than an antagonistic-class states. "Only a society that is able to establish a harmonious combination of productive forces according to a unified general plan can enable industry to be located around the country most conveniently for its development and also for the development of other elements of production.''[36]

Taking account of geographic conditions (natural and social) requires, along with knowledge of their individual elements, synthesized knowledge of the territorial complexes of the geographic environment (by countries, regions, and microregions) and of the interaction between its individual elements. Hence the inevitability of combining branch [systematic, Eds.] specialization with regional specialization in geography.

It is perfectly obvious that, in addition to branch-oriented geographers, life will require broad-based geographers with a regional specialization (physical geographers, economic geographers, regional geographers) who are able to do synthetic research and to compose works that give a comprehensive characterization of the natural and social conditions (geographic environment) of a certain country or region. The sooner the need for such synthetic geographers with a regional specialization is fully realized by the heads of planning and economic organizations, the sooner we shall arrive at a correct and comprehensive evaluation of geographic conditions, thereby riding practice of stereotyped management and ensuring the most rational location of our homeland's productive forces.

It does not occur to anyone today that it is necessary to prove the need for a specialist to manage a plant shop, a state farm, or collective farm. But in our country it is sometimes necessary to prove that the study of the geographic environment and its evaluation in practice must also be done by specialists. It is necessary to prove (even to geographers themselves) that, along with branch specialization, geography requires regional specialization and that knowledge of local characteristics and, consequently, their evaluation in practice are impossible without synthetic geographers who make a comprehensive study of an individual country (or a group of countries) and, with regard to large countries such as the USSR, the Chinese People's Republic, and the United States, their sections (regions).

Comprehensive knowledge of the geographic environment is totally indispensable to long-range planning and management of the economy of a country so large and diverse as our homeland, just as it is to the planning and management of the economies of its major parts. But this kind of knowledge can be obtained and correctly utilized only with the direct participation by geographers in the planning and management of the economy. Geographical research institutes alone are not enough here, although these institutions must also be created in all of the principal parts of our country, especially its eastern regions.

Broad-based geographers are still participating to a most inadequate degree in the practice of economic construction and probably even less in the practice of long-range planning. The opinion has taken hold that planning is the job only of economists and engineers.

Yet economics, despite all of its value, is not capable of equipping practitioners with concrete knowledge of the geographic environment. It is therefore not surprising that errors still occur in planning because of the planning agencies' poor knowledge of local charcteristics in the geographic environment.

Furthermore, the isolation of integrated geographers from practical work was promoted by the exclusively sector-oriented (departmental) organization of the management of the national economy. Under this kind of organization, specialists capable of making economic appraisals of territorial complexes of natural and economic conditions that have evolved differently in each region have seldom found a sphere of application for their efforts. The organization of regional economic councils [*sovnarkhozy*], as local bodies of territorial administration, can bring substantial improvement to this situation.

The miscomprehension of geography as a science has now caused an almost total discontinuance of training for geographers with an aptitude for synthetic works (broad-based geographers), that is, physical and economic geographers with a regional specialization, as well as regional specialists (who are especially needed for work in the geography of foreign countries) and has led to a *de facto* refusal to use such specialists in organizations that are in charge either of planning the national economy or of economic construction.[37]

The result has been a contradictory situation in which economic construction suffers in a number of instances from stereotyped management, caused by underestimation of local conditions, while the geography faculties of universities are curtailing the training of specialists capable of composing integrated characterizations of geographic conditions, because of the absence of a demand from economic organizations for such specialists.[38]

This situation is a natural cause for concern among university geographers. For example, A. M. Ryabchikov very rightly points out that ". . . the departmental separation of the components of geographical service among two dozen ministries and administrations and the absence of the position of geographical engineer on staff rolls make it most difficult to establish a firm liaison between university geography and production organizations. Meanwhile, the future of geography lies in these liaisons."[39] Indeed, the present organizational forms in which geographers can employ their knowledge to the needs of applied work are extremely unfavorable, and especially unfavorable for geographers with broad backgrounds.

Thus, questions of geographical theory that would seem to be very remote from concrete reality are actually of very great and direct practical importance. The correct solution of theoretical questions in geography can unquestionably help to eliminate many shortcomings that turn up in practice.

In summing up our examination of the question of regional geography, it should be noted first of all that we view regional geography as the part of geography that is concerned with the study of countries and regions defined by historical-geographical boundaries. At the same time it should be remembered that, whatever the scope of regional studies, it is limited. On the one hand, regional geography, as was pointed out earlier, excludes from its objects of study the earth's landscape envelope as a whole and many general geographical problems involving the study of the largest units of the earth's surface, continents and oceans. On the other hand, regional geography is not concerned with small regions that do not have historical-geographical boundaries, that is, it does not engage in local studies.

The part of geography that is concerned with individual types of landscapes, landscape units, natural zones, groups of enterprises, combinations of arable lands, territories adjacent to reservoirs, and so forth should be separated from regional geography and be given a *distinct* identity. Work in this kind of geography is of great and, one might say, the most direct importance to practice. This part of geography is beginning to develop its own techniques of inquiry and mapping and is establishing ties with related sciences. This distinct part of geography, in contrast to earth science and regional geography, may be called *local geography* or *topographic geography*.[40]

Local geography should enrich regional geography with factual material and provide typical examples of the most charcteristic features in the geography of individual countries and regions. Regional geography, in turn, should define in a number of instances the direction of local studies and generalize their results.

The major scientific and practical value and the characteristics of local geography as a part of geography (and not as a set of information about localities and regions) lie in the methods based on the large-scale approach of local studies. It would be wrong to view local studies as some ancillary, subsidiary division of regional studies.

The determination of local specificity in natural and economic conditions, the consideration of which is so important in the work of local agencies of economic management; the determination of local natural and economic conditions, the influence of which is so important to know when defining the economic specialization of administrative-economic regions; the uncovering of local resources for industrial and transport construction — all these (and many other) highly important *practical scientific* tasks can and must be carried out on the

basis of local studies. All of them define the practical orientation of local studies.

One cannot help but recall in passing the efforts to direct
the popular local-studies movement toward specific large-
scale geographical research of an applied-science character.
P. N. Stepanov, for example, wrote back in the 1930s: "...
When building a combine, we must do away with regional
'non-individuality,' we must apply as fully as possible the
principle of territorial utilization of the entire natural raw-
materials complex and the entire economic setting of the given
region. This means applying as fully as possible the *local-
studies* principle."[41] A broad local-studies movement chan-
neled through geography can really turn out to be the force
that will do away forever with stereotyped management of the
economy. Local studies as a part of geography, or local geography, is capable of properly equipping people in practice
with knowledge of specificity in local geographic conditions.

If one speaks of the distinctive characteristics in the approach to objects of inquiry that exist in local studies, one
must note first of all that local studies require expeditionary
and on-the-spot research. We are completely unable to im-
agine a compilatory local studies. It requires large-scale maps
or, more precisely, a good *topographic basis*; in a number
of cases local studies should make wide use of special
laboratories.

Thus, depending on the scale of inquiry, geography can be
divided into three major parts: *earth science, regional geography,* and *local geography.* Needless to say, there is a direct
intercausality between the development of these three parts of
geography. Earth science, regional geography, and local geography and are not distinct sciences but parts of geography.

However because geography is an integrated science composed of two basic groups of branches, the natural and the
social, regional geography (like earth science and local geography) is also integrated in character. Regional studies may
tend toward a more thorough presentation of the individual
branches of geography. It may be a regional-geographical
work, in which the country's nature, along with other ele-

ments, is presented most fully and thoroughly (as, for example, in the regional-geographical works of E. M. Murzayev), but it may also be a work that focuses its attention on the country's population, with wide use of data from history and ethnography. Moreover, as a general rule, each regional-geographical work will probably have some bias in favor of some group of geographical subfields, depending on the country, the purpose of inquiry, and the authors' specialization. One need not be afraid of this. It would be completely wrong to conclude from this that regional-geographical works are impossible, that it is impossible for regional studies to develop as a part of geography. But with the integrated character of geography, and therefore of regional geography, works on countries and regions are often divided into research in regional physical geography (which mostly provides a physical-geographical characterization of countries and regions) and research in regional economic geography (which mostly provides an economic-geographical characterization of countries and regions). In addition, there can be works in so-called specialized regional geography, which provide geographical material depending on the specific purpose of study of countries and regions (military, transport, agricultural, etc.).

In other words, to some extent regional geography cannot help but reflect the basic division of subfields in geography. But regional geography achieves the greatest integration, the greatest coherence between the branches of geography, and the greater the degree of coherence in a given inquiry, the more fully and correctly a country will be studied and depicted, and the more fully and better the regional-geographical work will be.[42]

Failure to understand this proposition leads sometimes to highly disparate interpretations as to what works may be called regional-geographical and what works may not. But it seems to us that any work of a broad geographical character, devoted to a given country or region defined by historical-geographical boundaries, will inescapably be regional-geographical. Research in the physical geography of countries and regions yields works in regional physical geography. Research in the geography of population and economy yields works in regional economic geography, and so forth.

It is also perfectly obvious that not all medium-scale geography (physical and economic) can come under regional studies, which under no circumstances can replace it completely. Regional studies by no means encompass (or substitute for) physical and economic geography, although they frequently unify them. Besides systematically oriented research, physical geographers can investigate the earth's landscape envelope, by discovering and studying physical geographic regions, whereas economic geographers can investigate the earth's landscape envelope by discovering and studying economic geographic regions. These broad works will not be regional-geographical works, since they interpret countries to be territories defined by state and historical-geographical boundaries.

One could cite a great many examples of geographical works (physical- and economic-geographical) that provide integrated characterizations of the geographic environment of countries and are therefore regional.[43] Unfortuantely, however, we still do not have a complete series of regional works on countries of the world or even on the countries that make up our Soviet Union, books that would be sufficiently comparable with each other. An exception, to some extent, is the series of books on the economic geography of the Union republics put out by the Geography State Publishing House ("The Blue Series"). The publication of these books may be considered an unquestionable achievement of Soviet geography; but the achievement is a relative one, since the books are written on very different levels.

But the production of comparable books on the geography of the countries of the world, written according to a basically unified plan, remains a feasible task for Soviet geographers, and the need for this kind of "Grand Geography of the World" is perfectly obvious.

The substantial differences that exist between countries and regions make it extremely difficult to compose a unified program for their regional characterization. An overly detailed program, if thoroughly carried out, is likely to do more harm than good. In addition to a general program outlining the basic, general features of regional characterizations and concerning itself with the comparability of the materials published in these characterizations (primarily statistical and cartographic), what are needed are separate, more detailed programs for each country individually. These specialized programs would focus attention on the geographic specificity of the country that distinguishes it from the other countries of the world and on presenting characteristic details that may be overlooked in a general program. Differences of this kind, which are inevitable when making up programs for regional characterizations of groups of countries, increase even more when specialized regional works are composed, when the goal of the regional characterization necessitates devoting more attention to some components of the geographic environment at the expense of others.

It would be totally wrong to demand of regional works that materials on all subfields of geography be provided in a strictly defined quantity, in a strict proportion, that they be measured "on a druggist's scale." Here geographers must remember *Lenin's thesis about the chief element,* must know how to find that chief element correctly and to discern and depict geographic specificity. And the chief element in regional works, depending on the character of the country and the purpose of its study, can be quite different things.

The purpose of the inquiry is of special importance in bringing out geographic specificity in specific regional works. For example, if a regional study of the Ukraine has the practical task of showing the complex of geographic conditions for the further development of agriculture, then the chief element of that study will be somewhat different than in a regional characterization of the Ukraine with the task of showing the complex of geographic conditions for the long-range planning of industrial and transport construction. In both cases good regional works on the Ukraine can be composed, but their orientation will be different, and hence the works themselves will be notably different.

The structure and content of regional works are also considerably influenced by the country's character. Regardless of the specific purpose of the inquiry, regional characterizations of territories so sharply diverse, say, as Greenland and the Caucasus, differ not only in factual material but in their entire structure.

In conclusion, we must return to the question of composing a "Grand Geography" of our homeland. The idea for this work was proposed by N. N. Baranskiy about thirty years ago; but although the idea always received the official support of most geographers, its implementation was initiated only in 1960.

To produce this kind of work, of course, is a difficult task. There are a whole host of obstacles. But it is also indisputable that the creation of a "Grand Geography" (both of the USSR and of the world) requires elimination of the dualistic views in geographical theory, because the locational definition of the subject matter of economic geography and the denial of the unity of geography as a science make it impossible to produce general geographical characterizations of countries and regions.

Geography as a whole must develop in our country as a science that synthesizes the results of all investigations of the elements of the landscape envelope. This, in our opinion, is the *chief task of geography as a whole,* and this is its basic *practical* role.

Because of the increase in demands from practice for the materials of geographic research, favorable conditions are being created for the development of geography.

Our country's entry into the period of full-scale construction of communism confronts Soviet geography with the necessity of investigating the whole complex of natural and economic conditions of specific territories with their economic appraisal and conclusions on the possibilities of fuller and more rational utilization in the process of production.

In this connection it is necessary to achieve better coordination of geographical writings and to create geography cells in institutions that are in charge of planning and managing the national economy. The problem of increasing the services of geographers for practical needs can be solved successfully only if there is a significant improvement in the organization of geographical research, especially of an applied-science character.

The best form of organization of Soviet geographers' work in the immediate future would be a *State Geographical Service* in charge of the study and conservation of geographic conditions and resources and also of their correct utilization. The creation of such a service would stimulate the study of the geographic environment, which in turn would produce even greater possibilities for the upsurge of productive forces. "If one estimated the extent of damage to the economies of all the countries of the world, including the USSR, that has come from inadequate or incorrect evaluation of local geographic conditions (natural and economic), one would find vast losses of labor that could have been avoided if there had been a geographical service capable of equipping practice with concrete knowledge of the geographic environment in terms of regional differences."[44]

The idea of creating a State Geographical Service, proposed by the geographers of the M. V. Lomonosov State University of Moscow, is most intimately connected with the monistic view of geography. Without a correct understanding of the unity of geography, a State Geographical Service is impossible.

In concluding the final chapter of this work, the following propositions must be noted:

1. The unity of geography's subject matter, in combination with the geographical method, makes it possible to inquire into the geographic environment not only by elements but also by their complexes, all the way up to the complete complex of all the elements in the geographic environment. This theoretically proves the possibility of autonomous integrated geographical research. Integrated studies of this kind, especially on small territories, can be highly diversified: they may be general geographical, physical-geographical, biogeographical,

economic-geographical, and specialized (applied-science) works. The combination of the elements studied in territorial complexes is determined in each case by the *purpose of inquiry*.

2. Geography is a science of territorial complexes. Consequently, by its very nature it can be at different scales. The geographical method, depending on the scale of inquiry, allows for the determination of territorial complexes with the most diverse degrees of generalization. The degree of generalization is of exceptional importance in geography.

Whereas the complex structure of the geographic environment gives rise to the *branch* differentiation of geography, the degree of generalization (scale of inquiry) gives rise to *its division into parts*. At the present stage of scientific development it is most expedient, obviously, to divide geography into three parts: *earth science, regional geography,* and *local geography.* Each of these parts, though not forming a distinct science, has important features peculiar to it. Therefore it is expedient for geography, along with its branch specializations, to specialize on the basis of its parts. In other words, the vastness of concrete material and its complexity demand *concrete regional specialization* for the integrated geographic approach.

Herein lies one of the important distinctions of the broad geographical approach as opposed to the branch approach, which does not have such an acute need for regional specialization. It is relatively easy for geomorphologists, climatologists, hydrologists, and so forth to change territory when they study their subject matter, which cannot be said about physical geographers, economic geographers, and especially regional geographers, since they deal not with individual elements but with complexes of elements of the geographic environment, which will have incomparably more differences from territory to territory than a member of any thematic geography will encounter. The complexes of the geographic environment, as a rule, are unique. Territories with very similar relief (or with a very similar climate) are easy to find; but territories of any size with very similar complexes of the geographic environment are impossible to find.

3. Geography is the science of the territorial complexes of the geographic environment. By determining and studying these complexes, it shows ways to make fuller and more rational use of them in practice.

By its very nature, geography is supposed to equip practitioners with knowledge of geographic conditions, knowledge of the specificity of these conditions from place to place. This is the basic practical significance both of geography as a whole and of any individual geographical science as a whole.

4. The practical importance of geography is becoming especially significant in the epoch of socialism, when the practical application of the results of geographical research is not hampered by the private-ownership mode of production.

In the conditions of the socialist socioeconomic system, which is directed toward maximum utilization of natural and social conditions and resources in the interests of mankind, unprecedented opportunities open up for the further subordination of nature to the interests of society.

Planned management of the economy has become possible under socialism, and because of this it has become especially important to have ample knowledge of the territorial complexes of the geographic environment. Without this knowledge, glaring errors in the long-range planning of the national economics of socialist countries are inevitable.

Taking account of territorial differences in the geographic environment rids applied work of stereotyped planning and promotes the correct integration of the branches of the national economy on a regional basis. But taking account of the geographic environment implies knowing it, and practitioners can be equipped with knowledge of the territorial complexes of the geographic environment only by an integrated science — *geography*.

The underestimation of geography in practice, which reflects a failure to understand its unity, has done and will continue to do noticeable harm to the national economy, by retarding its development and *lowering the productivity of labor*.

5. The successful implementation of practical-scientific, integrated geographical research at the present stage of de-

velopment of Soviet geography is connected in the closest way with the necessity of resolving a number of theoretical questions. The correct resolution of the theoretical question of the *unity* of geography is of especially great *practical* significance. This is based on the fact that the still widespread notion denying the unity of geography (and thereby denying the possibility of understanding the territorial complexes of the geographic environment) hampers the implementation of integrated, especially general, geographical research. The denial of the unity of geography theoretically justifies the exclusively thematic tendency in its development, and the elevation to an absolute of the general laws of development leads to a disregard of local peculiarities in natural and social conditons and to a stereotyped approach in economic practice that the Communist Party has denounced.

6. Elaboration of a theory of geography on the basis of Marxist-Leninist philosophy should be combined with more intensive development of integrated geographical works, above all those of an applied-science character.

In the process, the following trends should assume leading significance:

a. comprehensive study of the geographic environment of small areas for the purpose of satsifying the immediate needs of practice (the agricultural and silvicultural appraisal of lands, the study of water resources, the integrated study of regions with construction on the largest scale, etc.);

b. organization of effective control over correct utilization of the geographic environment in the process of production. Prevention of the squandering of natural resources;

c. determination and study of large economic-geographic regions. Elaboration of their characterizations, so that they may serve as an actual basis for long-range planning of the development of the economies of the USSR and the Union republics;

d. determination and characterization of economic-geographic regions (of the second order) with appraisal of their geographic conditions for the needs of regional economic councils;

e. development of integrated expeditions for the study of the deleterious consequences of irrational economic use of the geographic environment (erosion and so forth) with elaboration of a system of measures to reconstruct, preserve and enrich it;

f. codified geographical monographs in regional geography, and above all a *"Grand Geography of the USSR,"* and regional characterizations of foreign countries;

g. creation of public local-geographical mass organizations with the task of uncovering local resources and conditions for the needs of economic practice and the conservation of nature.

The above trends in the work of Soviet geographers, of course, are far from encompassing all the aspects of their study of the geographic environment. But these are the ones, in our view, that should be advanced at present as the principal ones, and special attention should be paid to them.

NOTES

CHAPTER ONE

1. K. Marx, *Das Kapital,* vol. 1 (Moscow, 1957), p. 57.

2. The Phoenician fleet was a rowing fleet propelled by slaves. The word *galera* [galley] is of Phoenician origin.

3. At any rate, the Phoenicians visited the Mediterranean countries and sold amber, which they brought from northern Europe.

4. Carl Ritter contends that the word *okean* [ocean] is of Phoenician origin (from the Phoenician *ogen,* which means "all-encompassing").

5. All these states had certain common cultural elements, and there was close interaction in their development; this gives grounds for referring to the Greek slave-owning society as a single entity.

6. The rise of Greece was aided by its geographical situation in the middle of the Eastern Mediterranean, that is, the center of the area of development of ancient culture. The high degree of fragmentation of the coast was also of great importance. In a number of places there are bays resembling tranquil lakes of a sort, with narrow outlets to the sea. From the time the sea became the chief means of communication between peoples, such a shoreline, which provided a large number of natural harbors, was highly conducive to the country's economic development. Greek maritime trade and its concomitant creation of Greek colonies in the coastal regions of the Mediterranean and the Black Sea were of great positive importance for the development of geography.

7. Geographical data may be found in Greek narrative literature, specifically in Homer's immortal works, the *Iliad* and the *Odyssey.* Although they provide vivid and colorful pictures of the ancient world, these poems at the same time attest to the contemporary dominance of mythological concepts of the earth as a whole.

8. For example, the problem of mutual causality in development between human society and the rest of nature.

9. One could cite many facts attesting to the development of astronomy and related roots of geography in prefeudal and feudal China. Ancient historical accounts (geography in China was closely tied to history) contain an enormous amount of material about which modern science still has a very poor knowledge.

10. G. V. Plekhanov, *K voprosu o razvitii monisticheskogo vzglyada na istoriyu* [On the Question of the Development of a Monistic View of History], in *Izbranniye filosofskiye proizvedenia* [Selected Philosophic Works], vol. 1 (Moscow, 1956), p. 615. Henceforth italics in quoted passages are those of the authors of the cited works, unless otherwise noted — V. A.

11. V. I. Lenin, *Konspekt knigi Gegelya "Lektsii po istorii filosofii"* [Abstract of Hegel's book, "Lectures on the History of Philosophy], in *Sochineniya* [Works], 4th ed., vol. 38, p. 247.

12. The establishment of a tie between society and nature may be credited to the materialist philosophers of the ancient world. Democritus, who was one of the first to tie the development of organic life to climatic conditions, wrote about the influence of the natural environment on man before Hippocrates.

13. True, Herodotus also considered climate to be the leading natural factor influencing the formation of the charcter of peoples.

14. At the same time, the polar and tropical zones were considered unlivable, an assumption that, it is true, was disputed by some scholars (Strabo and others).

15. The term "the earth's landscape envelope" was used for the first time by Yu. K. Yefremov in his article, *"O meste geomorfologii v kruge geograficheskikh nauk"* [On the Place of Geomorphology in the Field of Geographical Sciences], in *Voprosy geografii* [Questions of Geography], vol. 21 (Moscow, 1950), pp. 41–54.

16. *Istoria filosofii* [The History of Philosophy], vol. 1 (Moscow, 1957), p. 104.

CHAPTER TWO

1. *Istoria filosofii* [The History of Philosophy], vol. 1, p. 168.

2. From the Italian word *statista,* which means "person of the state."

3. Bernhard Varenius, *Geographia generalis in qua affectiones generales telluris explicantur* (Amsterdam, 1650). After the first publication in Amsterdam, Varenius's book was published in Cambridge, England, edited by I. Newton, in 1672 and again in 1681. The book was also published in other countries. Peter I selected Varenius's book as one of the first to be translated into Russian. Its first Russian edition: Varenius, *Geografia generalnaya, nebesny i zemnovodny krugi kupno i ikh svoistvy i deistvy v tryokh knigakh opisuyushchaya* [General Geography, the Celestial and Land-Water Circles Together, Their Properties and Motion Described in Three Books] (Moscow, 1718).

4. V. N. Tatischev, *O geografii voobsche i o Russkoi* [About General and Russian Geography], in *Izbranniye trudy po geografii Rossii* [Selected Works on the Geography of Russia] (Moscow, 1950), p. 210.

5. F. Engels, *Lyudvig Feierbakh i konets klassicheskoi nemetskoi filosofii* [Ludwig Feuerbach and the End of Classical German Philosophy], in K. Marx and F. Engels, *Izbranniye proizvedenia* [Selected Works] vol. 2 (Moscow, 1955), p. 341.

6. The concept of geographical determinism was elaborated a number of times long before Montesquieu. For example, as early as the sixteenth century Jean Bodin (1530–96) developed a theory about the influence of the geographical environment on the life of society. Montesquieu developed the same ideas, but his influence on the theoretical concepts of geographers was immeasurably greater than that of his predecessors.

7. C. Montesquieu, *O dukhe zakonov* [The Spirit of the Laws], in *Izbranniye proizvedenia* [Selected Works] (Moscow, 1955), p. 367.

8. Ibid.

9. Ibid., p. 412.

10. J. J. Rousseau, *Ob obshchestvennom dogovore* [The Social Contract] (Moscow, 1938), p. 67.

11. G. V. Plekhanov, *K shestidesyatoi godovshchine smerti Gegelya* [On the Sixtieth Anniversary of the Death of Hegel], in *Izbranniye filosofskiye proizvedenia* [Selected Works of Philosophy], vol. 1 (Moscow, 1956), p. 439.

12. C. A. Helvetius, *O cheloveke, yevo umstvennykh sposobnostyakh i vospitanii* [Of Man, His Mental Capacities and Upbringing] (Moscow, 1938), p. 83.

13. C. A. Helvetius, *Ob ume* [Of the Mind], (Moscow, 1938), p. 258.

14. D. Diderot, *Sistematicheskoye oproverzheniye proizvedenia Gelvetsia "Chelovek"* [A Systematic Rebuttal to the Work by Helvetius, "Man"] in *Sobraniye sochinenii* [Collected Works], vol. 2 (Moscow, 1935), p. 172.

15. N. G. Chernyshevsky, *Osnovania politicheskoi ekonomii* [The Foundations of Political Economy] in *Polnoye sobraniye sochinenii* [Complete Collected Works], vol. 9 (Moscow, 1949), p. 167.

16. P. Holbach, *Sistema prirody* [The System of Nature] (Moscow, 1940), p. 7.

17. Ibid., p. 20.

18. Ibid., pp. 21–22.

19. A. Humboldt, *Vozzrenia na prirodu* [Views of Nature], in *Magazin zemlevedenia i puteshestvy* [Journal of Earth Science and Travels], vol. 2 (Moscow, 1853), p. 2.

20. C. Ritter, *Obshcheye zemlevedeniye* [General Earth Science] (Moscow, 1864), p. 58.

21. C. Ritter, *Idei v sravnitelnom zemlevedenii* [Ideas in Comparative Earth Science], in *Magazin zemlevedenia i puteshestvy*, vol. 2 (Moscow, 1853), p. 375.

22. K. M. Ber, *O vliianii vneshnei prirody na sotsialniye otnoshenia otdelnykh naradov i istoriyu chelovechestva* [On the Influence of External Nature on the Social Relations of Individual Peoples and on the History of Mankind], in *Karmannaya knizhka dlya lyubitelei zemlevednia, izdavayemaya ot Russkovo geograficheskogo obshchestva* [Pocket Reader in Earth Science, Published by the Russian Geographical Society] (St. Petersburg, 1848), pp. 210, 230–31.

23. F. Ratzel, *Zemlya i zhizn* [The Earth and Man], vol. 2 (St. Petersburg, 1896), pp. 700–701.

24. É. Reclus, *Chelovek i zemlya* [Man and the Earth], vol. 1 (St. Petersburg, 1906), p. vii.

25. Ibid., p. 124.

26. L. Mechnikov, *Tsivilizatsia i velikiye istoricheskiye reki* [Civilization and Great Historical Rivers] (Moscow, 1924), p. 69.

27. Ibid., p. 162.

28. It was published only after the author's death, in French, edited and with a foreword by Élisée Reclus. Its only complete edition in Russian, issued in 1924, has long since become a bibliographical rarity; in addition, it is not without misprints and inaccuracies of translation. In effect, Russian readers have almost no opportunity to acquaint themselves with this interesting work by their compatriot; and if it is mentioned in our country, the purpose is merely to accuse it of geographical determinism.

29. G. V. Plekhanov, *O knige L. I. Mechnikova "Tsivilizatsia i velikiye istoricheskiye reki"* [On L. Mechnikov's Book, "Civilization and Great Historical Rivers"] in *Sochineniya* [Works], vol. 7 (Moscow-Petrograd, 1923), p. 28.

30. H. Buckle, *Istoria tsivilizatsii v Anglii* [The History of Civilization in England], vol. 1 (St. Petersburg, 1895), p. 16.

31. I. A. Vitver, *Frantsuzskaya shkola "geografii cheloveka"* [The French School of "Human Geography"], in *Ucheniye zapiski MGU* [Scholarly Notes of Moscow State University], no. 35 (Moscow, 1940), p. 41.

32. É. Reclus, *Chelovek i zemlya*, vol. 1, p. 113.

33. Ibid., p. 115.

34. Ibid., p. x.

35. The most thorough analysis of the works of French geographers is contained in I. A. Vitver's work, "The French School of Human Geography." This work is supplemented by a small but meaty article by S. F. Biske, *"O frantsuzskoi geograficheskoi shkole i frantsuzskoi geograficheskoi literature za 1940–46 gody"* [On the French Geographical School and French Geographical Literature 1940–1946], in *Izvestia VGO* [News of the All-Union Geographical Society], no. 3 (1947), pp. 335–41. Some volumes of *Geographie Universelle* have been translated into Russian and published in the USSR.

36. "Three conditions have a particular influence on the life of the people: the nature of the country in which they live, the nature of the tribe to which they belong,

the course of external events and influences coming from peoples that surround them." S. M. Solovyov, *Nachalo russkoi zemli* [The Origin of the Russian Land], in *Sbornik gosudarstvennykh znany* [Collection of the State's Knowledge], vol. 4 (St. Petersburg, 1877), p. 1. "Various tribes and peoples still are, in many respects, living imprints of the natural forms and types of the localities where they are born and cultivated" A. P. Shchapov, *Sochineniya* [Works], vol. 2 (St. Petersburg, 1906), p. 173. "A state's strength, having been based in the region of the sources of a plain's major rivers, naturally strove to extend the sphere of its dominion to their mouths . . . . Thus, the center of a state's territory was determined by the upper reaches of rivers; its circumference, by their mouths; and its further settlement, by the direction of the river basins. This time our history has proceeded in close agreement with natural conditions: rivers, to a large extent, outlined its program" V. O. Klyuchevsky. *Kurs po russkoi istorii* [Course in Russian History], part 1, in *Sochineniya*, vol. 1 (Moscow, 1956), p. 65. Klyuchevsky has a whole host of propositions in which he links the distinctive psychological and national characteristics of the Russian people to the characteristics of the natural environment. Despite their excessive categoricalness and, therefore, inaccuracy, these propositions contain a grain of reason, since a people's national charcteristics undoubtedly evolved, among other things, under the influence of the surrounding nature.

37. N. V. Gogol, *Vzglyad no sostavleniye Malorossii* [A View of the Formation of Little Russia], in *Sobraniye sochineny v shesti tomakh* [Collected Works in Six Volumes], vol. 6 (Moscow, 1953), p. 27.

38. B. B. Polynov, *Ucheniye o landshaftakh* [Theory of Landscapes], in *Izbranniye trudy* [Selected Works] (Moscow, 1956), p. 492.

39. I. M. Zabelin, *Osnovniye problemy teorii fizicheskoi geografii* [Basic Problems of the Theory of Physical Geography] (Moscow, 1957), p. 8.

CHAPTER THREE

1. V. I. Lenin, *Materializm i empiriokrititsizm* [Materialism and Empirio-criticism], in *Sochineniya* [Works] 4th ed. vol. 14, pp. 184–185.

2. I. Kant, *Magistra Immanuila Kanta programma raspredelenia lektsy v zimnem polugodii 1756–1766 gg.* [Master Immanuel Kant's Program for the Distribution of Lectures in the Winter Semester, 1765–1766], in *Sochineniya* [Works], vol. 2 (Moscow, 1940), p. 312.

3. I. Kant, *O forme i printsipakh chuvstvennogo i umopostigayemogo mira* [On the Form and Principles of the Sensible and Intelligible World], in *Sochineniya*, vol. 2, p. 410.

4. Ibid., p. 413.

5. I. Kant, *O razlichnykh rasakh lyudei* [Concerning the Various Races of People], in *Sochineniya*, vol. 2, p. 463.

6. V. I. Lenin subjected Hume's indeterminist philosophy to devastating criticism in his work *Materializm i empiriokrititsizm*.

7. H. Rickert, *Nauki o prirode i nauki o kulture* [Sciences of Nature and Sciences of Culture] (St. Petersburg, 1911).

8. True, Rickert spoke of the natural-historical method of gaining knowledge, which sometimes provided grounds for seeing elements of determinism in his concepts. In reality, however, Rickert's determinism was particularly nominal. The materialistic nature of determinism was profoundly alien to Rickert, and his so-called natural-historical method was nothing more than a Kantian category of purely formal logic that had nothing in common with the authentically natural-historical method that was developed in the works of many naturalists.

9. N. N. Kolosovsky, *Nauchniye problemy geografii* [Scientific Problems of Geography], in *Voprosy geografii* [Questions of Geography], no. 37, p. 131.

10. N. N. Baranskiy, *Kratky ocherk razvitia ekonomicheskoi geografii* [Brief Account of the Development of Economic Geography], in the collection *Ekonomicheskaya geografia. Ekonomicheskaya kartografia* [Economic Geography and Economic Cartography] (Moscow, 1960), p. 22.

11. F. Engels, *Ludvig Feierbakh i konets klassicheskoi nemetskoi filosofii* [Ludwig Feuerbach and the End of Classical German Philosophy], in K. Marx and F. Engels, *Izbranniye proizvedenia* [Selected Works], vol. 2, pp. 354–55.

12. Hegel, *Sochineniya* [Works], vol. 8 (Moscow, 1935), p. 76.

13. Ibid., vol. 2, (Moscow, 1930), p. 342.

14. Ibid., vol. 8, p. 76.

15. This kind of specifically idealistical appraisal of the theoretical concepts of Humboldt and Ritter is especially widespread in foreign geography. For example, Richart Hartshorne's well-known work, *The Nature of Geography,* which is undoubtedly the most thorough work on the theory of geography to come out in the United States, considers the views of these two outstanding geographers of the past from precisely this standpoint. It was a synthesis of the works of Humboldt and Ritter that gave geography its present form, Hartshorne asserts. In addition, Humboldt and Ritter are declared the creators of geography as a science. One of the chapters in Hartshorne's book has just such a title: "The Classical Period: Humboldt and Ritter" (see R. Hartshorne, *The Nature of Geography* [New York, 1939]). In fact, Hartshorne goes even further. In one of his latest works — "The Concept of Geography as a Science of Space, from Kant and Humboldt to Hettner" — he seeks to establish complete identity in the methodological views of Kant, Humboldt and Hettner on the nature of geography and its place in the classification of sciences. (See R. Hartshorne, "The Concept of Geography as Science of Space, From Kant and Humboldt to Hettner," in *Annals, Association of American Geography,* 1958, no. 2.) Yet even representatives of Russian prerevolutionary geography stressed the fundamentally important differences between the works of A. Humboldt and C. Ritter. For example, I. V. Mushketov wrote: "But the tremendous difference between them consists in the fact that Humboldt based his conclusions on personal observations of nature and used sources with great selectivity, being able to take a strict, critical attitude toward them, since he was a leading expert for his time in many branches of knowledge, especially physical geography and geology; in this respect he was the complete opposite of Ritter, who discussed nature almost all his life while looking at it from the window of his study.

"In addition to brilliant generalizations, Humboldt contributed a mass of new facts to science. Ritter only systematized what already existed, illuminating it with a certain concept. Humboldt considered the diverse phenomena of nature in terms of their internal relations and tried to ascertain their genesis; Ritter based everything on plastic movements, on external configuration, without trying to explain the genesis of homologous forms. Humboldt acknowledged the influence of nature on man, but did not try to construct an autonomous, separate science on this maxim; he regarded the earth not only as a physical body but also as a world body, and he sought to extend and comprehend the general ideas about the world edifice. Ritter, meanwhile, wanted to create on this basic maxim [the influence of nature on man — V. A.] a completely new but impossible science." I. V. Mushketov, *Turkestan,* vol. 1 (Petersburg, 1915), p. 137.

16. F. Engels. *Polozheniye Anglii. Vosemnadtsaty vek* [The Situation of England in the Eighteenth Century], in K. Marx and F. Engels, *Sochineniya* [Works], 2d ed., vol. 1, p. 599.

17. It was not by mere chance that A. Humboldt began his principal synthetic work, *The Cosmos,* at the age of seventy-five.

18. A. Humboldt, *Kosmos,* part 1 (Moscow, 1866), p. iii.

19. Ibid., p. 62.

20. Ibid., p. 11.

21. Ibid., p. 31.

22. Ibid., p. 53.

23. Ibid., p. 39.

24. Ibid., p. 261.

25. D. N. Anuchin, *Aleksandr fon-Gumboldt kak puteshestvennik i geograf i v osobennosti kak issledovatel* [Alexander Von Humboldt as a Traveler and Geographer and Particularly as a Researcher], in *Geograficheskiye raboty* [Geographical Works] (Moscow, 1954), p. 365.

26. I. P. Gerasimov, *Sostoyaniye i zadachi sovetskoi geografii na sovremennom etape yeyo razvita* [The State and Tasks of Soviet Geography at the Present Stage of Its Development], in *Izvestia AN SSSR* [News of the USSR Academy of Sciences], geography series, 1955, no. 3.

27. J. Herder, *Mysli, otnosyashchiyesya k filosofskoi istorii chelovechestva* [Thoughts About the Philosophical History of Mankind] (St. Petersburg, 1829), p. 29.

28. Herder had a considerable and, for the most part, propitious influence on his friend, the great German poet Goethe.

29. Hegel, *Filosofia prirody* [Philosophy of Nature], in *Sochineniya* [Works], vol. 2, pp. 357–58.

30. C. Ritter, *Idei o sravnitelnom zemlevedenii* [Ideas About Comparative Earth Science], in *Magazin zemlevedenia i puteshestvi* [Journal of Earth Science and Travels], vol 2, p. 416.

31. I. V. Mushketov, *Turkestan,* vol. 1, p. 131.

32. C. Ritter, *Idei o sravnitelnom zemlevedenii,* in *Magazin zemlevedenia i puteshestvy,* vol. 2, p. 363.

33. D. N. Anuchin, *O prepodavanii geografii i o voprosakh, s nim svyazannykh* [On the Teaching of Geography and Related Questions] in *Geograficheskiye raboty* [Geographical Works], p. 295.

34. Yu. G. Saushkin, *Vvedeniye v ekonomicheskuyu geografiyu* [Introduction to Economic Geography] (Moscow, 1958), p. 77.

35. By that time the theoretical meagerness of the concepts of geographical determinism had been demonstrated quite amply.

36. The development of anthropogeography, which viewed human society only as a biological phenomenon on earth, essentially was also a kind of retreat from questions that seemed insoluble. There are instances in science in which incomprehension has led to denials. More specifically, the failure to comprehend the qualitative distinctiveness of human society and the specific character of the laws of social development led to a tendency to account for social phenomena by natural laws. Today we sometimes also encounter a denial of phenomena that seem obscure. Thus, although they know particular laws of social development and see the qualitative distinctiveness of society by comparison with the rest of nature, some present-day geographers and economists infer on this basis absolute opposition between society and the rest of nature. We have here two sides of the same coin. In both instances we encounter a denial that is due to a failure to comprehend. In the first instance the specificity of the particular was not understood; in the second, the universality of the whole was not understood.

37. I. V. Michurin, *Otvety na voprosy redaktsii zhurnala "Za marksistskoleninskoye yestestvoznaniye"* [Answers to Questions From the Magazine "For a Marxist-Leninist Natural Science"], in *Sochineniya* [Works], vol. 1 (Moscow, 1948), p. 623.

38. The proponents of such an attitude toward the theoretical heritage left to us by the bourgeois geographers of the past usually do not take the trouble to study their views, restricting themselves to an assessment by the method of spouting labels that have long since become hackneyed.

39. F. Ratzel, *Chelovechestvo kak zhiznennoye yavleniye na zemle* [Mankind as a Phenomenon of Life on Earth] (Moscow, 1901), p. 84.

40. F. Ratzel, *Zemlya i zhizn* [The Earth and Life], vol. 2, p. 670.

41. Ye. Tarle, one of the most outstanding Soviet historians, gave this definition of geopolitics: "... Geopolitics is the doctrine of why present-day German fascism wishes to snatch from its neighbors certain territories, which of them are to be snatched first, which second, and of how to prepare most adeptly and expediently the ideological groundwork and favorable atmosphere for the successful display of this 'racial will' to rob neighbors." Ye. Tarle, *Vostochnoye prostranstvo i fashistskaya geopolitika* [The Eastern Territories and Fascist Geopolitics], in the collection *Protiv fashistskoi falsifikatsii istorii* [Against the Fascist Falsification of History] (Moscow, 1939), p. 270. It is perfectly obvious that the validity of this definition is not changed by the national origin of a given fascism.

42. A. Strahler, *Physical Geography* (New York, 1951).

43. F. Engels, *Pismo k Danielsonu* [Letter to Danielson], in K. Marx and F. Engels, *Sochineniya* [Works], vol. 28, p. 56.

44. C. Haskins, *Of Societies and Men* (New York, 1951).

45. E. Huntington, *Mainsprings of Civilization* (New York, 1945).

46. *Geography in the Twentieth Century* (New York and London, 1951).

47. R. S. Platt, "Determinism in Geography," in *Annals, Association of American Geographers*, 1948, no. 2.

48. A. Hettner, *Geografia. Yeyo istoria, sushchnost i metody* [Geography: Its History, Character and Methods] (Moscow, 1930), p. 115. [Translation of Alfred Hettner, *Die Geographie, ihre Geschicte, ihr wesen und ihre Methoden* (Breslau, 1927).]

49. In our country a critical, scientific analysis of the works of foreign geographers (A. Hettner in particular) has, on occasion, been forgone in favor of sweeping vilification. An example of this is the review of the collection, *Burzhuaznaya geografia na sluzhbe amerikanskogo imperializma* [Bourgeois Geography at the Service of American Imperialism] (Moscow, 1951), published in the magazine *Voprosy ekonomiki* [Problems of Economics], 1951, no. 2.

50. A. Hettner, *Geografia. Yeyo istoria, sushchnost i metody*, p. 116.

51. Ibid.

52. Ibid.

53. Ibid., p. 197.

CHAPTER FOUR

1. V. P. Semenov-Tyan-Shansky, *Rol gosudarstvennogo Tsentralnogo geograficheskogo muzeya v dele novogo prepodavania geografii* [The Role of the State Central Geographical Museum in the New Teaching of Geography], in *Geografia v shkole* [Geography in School], no. 2 (1934), p. 33.

2. F. Engels, *Razvitiye sotsializma ot utopii k nauke* [The Development of Socialism From Utopia to Science], in K. Marx and F. Engels, *Izbranniye proizvedenia* [Selected Works], vol. 2, p. 120.

3. In this respect V. V. Dokuchayev was not alone in Russian science: A. I. Voyeikov, L. I. Prasolov, B. B. Polynov, and V. N. Sukachov also arrived at an

understanding of the unity of the earth's landscape envelope after studying the individual components of nature.

4. V. V. Dokuchayev, *Mesto i rol sovremennogo pochvovedenia v nauke i zhizni* [The Place and Role of Modern Soil Science in Science and Life], in *Sochineniya* [Works], vol. 6 (Moscow-Leningrad, 1951), pp. 416–17.

5. However, the works of L. S. Berg, as well as those of many Soviet landscape scientists, have a mistaken inclination to restrict their study of landscapes to the complexes of elements that formed exclusively under the effect of natural laws. Later we shall return to this question, especially since L. S. Berg made references to geography as a science of "the filling of space," that is, references that are essentially close to A. Hettner's theory, which we have already examined. But the work itself with its specific material has obviously led and is leading landscape scientists to recognize, if not in word, then in fact, the unity of the natural and social on the face of the earth, in the structure of landscapes.

6. V. V. Dokuchayev, *K ucheniyu o zonakh prirody* [On the Theory of Natural Zones], in *Sochineniya* [Works], vol. 6, pp. 399.

7. V. V. Dokuchayev, *Sochineniya,* vol. 6, pp. 375–77.

8. V. V. Dokuchayev, *Mesto i rol sovremennogo pochvovedenia v nauke i zhizni* [The Place and Role of Modern Soil Science in Science and Life], in *Sochineniya,* vol. 6, p. 421.

9. B. B. Polynov, *Razvitiye idei Dokuchayeva v zapadnoyevropeiskoi nauchnoi literature* [The Development of Dokuchayev's Ideas in Western European Literature], in *Izbranniye trudy* [Selected Works] (Moscow, 1956), p. 606.

10. D. N. Anuchin, *Geografia (iz ents. sl. Granat)* [Geography (From Granat's Encyclopedic Dictionary)], in *Geograficheskiye raboty* [Geographical Works] (Moscow, 1954), pp. 313–14.

11. Ibid., p. 318.

12. A. A. Grigoryev, *Znacheniye idei i nauchnogo tvorchestva D. N. Anuchina* [The Value of the Ideas and Scientific Works of D. N. Anuchin], in D. N. Anuchin, *Geograficheskiye raboty,* p. 9.

13. D. N. Anuchin, *O prepodavanii geografii i o voprosakh, s nim svyazannykh* [On the Teaching of Geography and Related Questions], in *Geograficheskiye raboty,* p. 296.

14. This raises the possibility of misinterpretation of their concepts. Moreover, it should not be forgotten that neither one was a Marxist, and as a result they were unable in a whole host of cases to find correct solutions to basic problems of geography.

15. Another difficulty in studying the relations between individual components and the whole is that, in singling out the components, we seem to destroy or lose sight of the whole. But in studying the whole without singling out its constituent parts, we cannot comprehend this whole in all of its details, and our image of it will only be general and not specific. For this reason alone, analysis and synthesis taken separately are not capable of assuring a process of scientific cognition.

16. A. M. Ryabchikov, *O putyakh razvitia universitetskoi geografii* [On the Paths of Development of University Geography], in *Vestnik MGU* [Moscow University Herald], series 5, 1960, no. 1, p. 7.

17. V. I. Lenin, *Konspekt knigi Aristotelya "Metafizika"* [Critical Remarks on Aristotle's Book, "Metaphysics"], in *Sochineniya* [Works], 4th ed., vol. 38, p. 370.

18. K. Marx and F. Engels, *Nayemny trud i kapital* [Hired Labor and Capital], in *Sochineniya,* 2d ed., vol. 6, p. 441.

19. D. L. Armand, *Predmet, zadacha i tsel fizicheskoi geografii* [The Subject

Matter, Task and Purpose of Physical Geography], in *Voprosy geografii* [Questions of Geography], no. 40 (Moscow, 1957), pp. 68–102.

20. F. Engels, *Dialektika prirody* [The Dialectics of Nature] (Moscow, 1955), p. 16.

21. Ibid., p. 191.

22. V. I. Lenin, *Konspekt knigi Gegelya "Nauka logiki"* [Critical Remarks on Hegel's Book, "The Science of Logic"], in *Sochineniya,* 4th ed., vol. 38, p. 140.

23. N. N. Baranskiy, *Ocherki shkolnoi metodiki ekonomicheskoi geografii* [Essays on School Methods in Economic Geography] (Moscow, 1955), p. 148.

24. It was literally under this maxim that the article on geography, for example, was written in the second edition of the *Great Soviet Encyclopedia* (vol. 10, pp. 456–68). About geography as a whole the article says nothing; it has been eliminated; the authors of the encyclopedia article obviously recognize the existence only of individual, completely autonomous geographies.

25. F. Engels, *Dialektika prirody,* pp. 180–81.

26. I. S. Shchukin, *O meste geomorfologii v sisteme yestestvennykh nauk i o yeyo vzaimootnosheniakh s kompleksnoi fizicheskoi geografiyei* [On the Place of Geomorphology in the System of Natural Sciences and Its Interrelations with Integrated Physical Geography], in *Vestnik MGU,* series 5, 1960, no. 1, p. 19.

27. Ibid.

28. I. P. Gerasimov, *Rol geografii v sotsialisticheskom stroitelstve SSSR i sovremenniye tendentsii yeyo razvitia* [The Role of Geography in the Socialist Construction of the USSR and Present Trends in its Development], in *Voprosy geografii* [A collection of articles for the Eighteenth International Geographical Congress] (Moscow-Leningrad, 1956), p. 8.

29. Thor Heyerdahl, *Aku-Aku, taina ostrova Paskhi* [Aku-Aku: the Secret of Easter Island], in *Yunost* [Youth], 1958, no. 3, p. 72.

30. N. N. Baranskiy, *Stranovedeniye i geografia fizicheskaya i ekonomicheskaya* [Regional, Physical and Economic Geography], in *Ekonomicheskaya geografia. Ekonomicheskaya kartografia* [Economic Geography and Economic Cartography], p. 161.

31. V. N. Sementovsky, *O profile fiziko-geografa* [On the Background of a Physical Geographer], in *Voprosy geografii,* no. 25 (Moscow, 1951), p. 71.

32. Ibid., pp. 71–72.

33. Ibid., p. 72.

34. Ibid., p. 60.

35. N. N. Kolosovsky, *Nauchniye problemy geografii* [Scientific Problems of Geography], in *Voprosy geografii,* no. 37 (Moscow, 1955), p. 145.

36. N. N. Baranskiy repeatedly came out against the obliteration of geography by means of its overspecialization. See, for example, his article: "Ob uchebnykh planakh i ob obshchegeograficheskom otdelenii na geograficheskom fakultete" [On Academic Programs and the General Geographical Section in the Geography Faculty], in *Ekonomicheskaya geografia v srednei shkole. Ekonomicheskaya geografia v vysshei shkole* [Economic Geography in Secondary School and in Higher Education] (Moscow, 1957), pp. 256–77.

37. P. James and C. Jones, eds., *Amerikanskaya geografia* [American Geography] (Moscow, 1957), p. 26 [p. 4 in original]. The quotations are from the Russian-language edition; the book was published in English in the U.S.A. in 1954 — V. A. [Translation of P. E. Jones and C. F. Jones, eds., *American Geography: Inventory and Prospect* (Syracuse, N.Y.: Syracuse University Press, 1954).]

38. Ibid., pp. 34–35 [p. 15].

39. See, for example, the article by D. Hooson, "Some Recent Developments in the Content and Theory of Soviet Geography," in *Annals, Association of American Geographers,* vol. 49, no. 1 (1959), pp. 73–82.

40. P. James and C. Jones, *Amerikanskaya geografia,* p. 59 [p. 44].

41. Ibid., p. 9 (N. N. Baranskiy's foreward).

42. I. M. Zabelin, *Teoria fizicheskoi geografii* [The Theory of Physical Geography] (Moscow, 1959), p. 37.

43. Ibid., p. 18.

CHAPTER FIVE

1. Yu. K. Yefremov, *Landshaftnaya sfera Zemli* [The Landscape Sphere of the Earth], in *Izvestia VGO* [News of the All-Union Geographical Society], 1959, no. 6, p. 528.

2. Ibid., p. 525.

3. True, here as well there are some differences between the landscape envelope and the geographic environment. Certain natural elements of the landscape envelope even on inhabited territory do not belong to the geographic environment (e.g., the upper strata of the atmosphere). On the other hand, in the process of production mankind can use resources of the earth that are located outside of the landscape envelope.

But for all intents and purposes, this difference is of no great importance. All the elements of the landscape envelope on the inhabited part of the earth's surface gradually become part of the geograpic environment, and usually it is just from this time on that they become objects of geographical study. And the resources that are extracted from the depths of the earth, beyond the landscape envelope, are utilized (i.e., are in essence turned into an element of the environment) inside of it.

4. K. Marx, *Das* Kapital, vol. 1, p. 184.

5. F. Engels, *Dialektika priorody* [The Dialectics of Nature], p. 141.

6. F. Engels, *Lyudvig Feierbakh i konets klassicheskoi nemetskoi filosofii* [Ludwig Feuerbach and the End of Classical German Philosophy], in K. Marx and F. Engels, *Izbranniye proizvedenia* [Selected Works], vol. 2, p. 371.

7. G. V. Plekhanov, *K voprosu o razvitii monisticheskogo vzglyada na istoriyu* [On the Question of the Development of the Monistic View of History], in *Izbranniye filosofskiye proizvedenia* [Selected Works of Philosophy], vol. 1, (Moscow, 1956), p. 608.

8. Ibid., p. 615.

9. The highly diverse social functions that people perform in the process of production are simultaneously various forms of their organisms' vital activity. Any labor is an expenditure of energy associated with the work of human muscles and the human brain. Therefore the character of labor, which is determined by the social system, the level of technology, and the influence of the geographic environment, in turn determines the development of both the geographic environment as a whole and of man himself within this environment. Therefore the character of labor in widely disparate geographic conditions will also be somewhat disparate; in countries with extreme differences in the geographic environment, obviously there will always be considerable differences in the population, even under a totally identical social system.

10. Furthermore, the process of making use of more and more elements of nature for production is a social process that is realized through the utilization of the laws of nature (with the aid of technology). It should not be forgotten that both producers and

the means of production are reproduced according to the laws of nature, the effect of which is directed by the mode of production. But under any mode of production, no development or reproduction of productive forces and no development of technology is possible without the utilization of natural laws.

11. V. I. Lenin, *Materializm i empiriokrititsizm* [Materialism and Empirio-criticism], in *Sochineniya* [Works], 4th ed., vol. 14, p. 156.

12. V. I. Lenin, *Konspekt knigi Geglya "Nauka logiki"* [Critical Remarks on Hegel's Book, "The Science of Logic"], in *Sochineniya*, 4th ed., vol. 38, p. 204.

13. V. I. Lenin, *Materializm i empiriokrititsizm*, in *Sochineniya*, 4th ed., vol. 14, p. 81.

14. N. N. Kolosovsky, *Nauchniye problemy geografii* [Scientific Problems of Geography], in *Voprosy geografii* [Questions of Geography], no. 37 (Moscow, 1955), p. 132.

15. K. Marx, *Das Kapital*, vol. 2, p. 184.

16. Ibid., p. 191.

17. V. I. Lenin, *Materializm i empiriokrititsizm*, in *Sochineniya*, 4th ed., vol. 14, p. 143.

18. F. Engels, *Razvitiye sotsializma ot utopii k nauke* [The Development of Socialism From Utopia to Science], in K. Marx and F. Engels, *Izbranniye proizvedenia* [Selected Works], vol. 2, p. 121.

19. V. I. Lenin, *Agrarny vopros i "kritiki" Marksa* [The Agrarian Question and Marx's "Critics"], in *Sochineniya*, 5th ed., vol. 5, p. 103.

20. F. Engels, *Anti-Dyuring* [Anti-Dühring], (Moscow, 1957), p. 107.

21. Ibid.

22. F. Engels, *Dialektika prirody*, p. 138.

23. F. Engels, *Ob avtoritete* [On Authority], in K. Marx and F. Engels, *Izbranniye proizvedenia*, vol. 1, pp. 589–90.

24. F. Engels, *Dialektika prirody*, p. 14.

25. K. Marx, *Das Kapital*, vol. 1, p. 188.

26. It is completely clear that the geographic environment is a correlative concept. It loses meaning whenever we are dealing with the study of dehumanized nature or nature is viewed as such without any consideration of its role in social life.

27. F. Engels, *Dialektika prirody*, p. 183.

28. A. Ye. Fersman, *Geokhimia* [Geochemistry], vol. 2 (Leningrad, 1934), p. 297.

29. K. K. Markov, *Paleogeografia* (Moscow, 1951), p. 8.

30. A. M. Ryabchikov, *O putyakh razvitia universitetskoi geografii* [On the Paths of Development of University Geography], in *Vestnik MGU* [Moscow University Herald], series 5, 1960, no. 1, p. 7.

31. Yu. G. Saushkin, *Vvedeniye v ekonomicheskuyu geografiyu* [Introduction to Economic Geography], (Moscow, 1958), p. 22.

32. I. S. Shchukin, *O meste geomorfologii v sisteme yestestvennykh nauk i o yeyo vzaimootnosheniyakh s kompleksnoi fizicheskoi geografiyei* [On the Place of Geomorphology in the System of Natural Sciences and Its Interrelations with Integrated Physical Geography], in *Vestnik MGU*, series 5, 1960, no. 1, p. 16.

33. Ibid., p. 17.

34. The best definition by a Soviet geographer of the object of study and the nature of geography was given, in our view, by Yu. G. Saushkin: "Thus, physical geography and economic geography examine on the face of the earth their own material objects of study, which exist in reality: The first are natural complexes (zones,

regions, landscapes), the second, production complexes. These material objects of study take shape on a certain territory, have boundaries, have followed a certain historical course of development, have their own futures and interact with each other. No other sciences but the geographical sciences study *territorial* complexes. Consequently, the task of the geographical sciences, and only the geographical sciences, is to investigate the various kinds of territorial complexes that have evolved on the face of the earth, including their history and their geographic extent, that is, to investigate them in terms of both time and space, with emphasis on pointing out spatial patterns." See Yu. G. Saushkin, *Vvedeniye v ekonomicheskuyu geografiyu* [Introduction to Economic Geography], p. 18.

35. N. N. Kolosovsky, *Nauchniye problemy geografii* [Scientific Problems of Geography], in *Voprosy geografii* [Questions of Geography], no. 37, p. 139.

36. K. Marx, *Das Kapital,* vol. 2, (Moscow, 1955), p. 358.

37. W. R. Williams, *Lenin o plodorodii* [Lenin on Fertility], in *Kolkhoznik* [Collective Farmer], 1939, no. 2, p. 98.

38. Haiyuntsang were food storehouses on the bank of an ancient canal in the eastern part of Peking and the original location of the People's University of China after its transfer to Peking.

39. Sun Ching-Chi, *Ekonomicheskaya geografia kak nauka* [Economic Geography as a Science], Moscow, 1959, p. 38.

40. L. S. Berg, *Geograficheskiye zony Sovetskogo Soyuza* [Geographic Zones of the Soviet Union], vol. 1 (Moscow, 1947), p. 23.

41. N. N. Kolosovsky, *Nauchniye problemy geografii*, in *Voprosy geografii*, no. 37, p. 142.

42. B. M. Kedrov, *O klassifikatsii nauk* [On the Classification of Sciences], in *Voprosy filosofii* [Questions of Philosophy], 1955, no. 2, pp. 49–68.

43. V. P. Semenov-Tyan-Shansky, *Raion i strana* [Region and Country] (Moscow-Leningrad, 1928), p. 45.

44. One important regularity should be recalled here. The laws of the development of less perfect forms of matter do not disappear in its more perfect forms, although the laws of development of these latter do not extend to less perfect forms. Therefore, although the laws of development of human society are fundamentally different from the laws of development of the rest of nature, these opposite forms of the material world retain a definite commonality, which is eternal and never disappears. The laws of physics and chemistry that operate in the atmosphere do not cease to operate in the biosphere. The laws of physics, chemistry, and biology that operate in the biosphere do not cease to operate in human society (but this does not apply to social relations), although they, of course, are not what determines social development.

45. For example, Great Britain's island situation resulted from purely natural factors. But isn't it important to an understanding of the geography of that country's population and economy? If Great Britain were a peninsula, its present population and economy would be completely different, just as its history would have been somewhat altered.

46. I. M. Zabelin, *Teoria fizicheskoi geografii* [A Theory of Physical Geography], p. 272.

47. The nature of the difference between the two close but not identical concepts, "object of study" and "subject matter," may be clarified in an example.
The geographic environment is the subject matter of geography. The subject matter of physical geography is the natural complex of elements of the geographic environment, whereas the entire geographic environment is the object of study of physical geography, because when physical geographers make a specialized study of the natural complex, they study it as a part of the geographic environment as a whole, in

connection with the general laws of its development and in connection with the other elements that it comprises. The subject matter for geomorphology is relief, and the entire natural complex of the geographic environment (as well as the geographic environment as a whole) will be the object of study. The object of study is a more general concept. Subject matter is a more specific concept (for a given science). In broad terms, it may be said that matter is the subject matter of science as a whole and at the same time is the object of study for all the sciences.

48. Geography is at once a system of sciences concerned with individual elements of the geographic environment and an integrated science, since it is capable of understanding the geographic environment as a whole. This may be clarified more concretely in the case of physical geography, which, on the one hand, is also a system of geographical sciences concerned with the natural elements of the geographic environment and, on the other hand, it is a science capable of comprehending integrated combinations of these elements and is capable of autonomous inquiry into these combinations. An example of such synthetic physical-geographical research is work in landscape science. The integrated character of the object of study inevitably leads to the integration of the science concerned with this object of study. However, there are essentially no nonintegrated sciences, just as there is no whole that cannot be divided into parts. The lack of fragmentation of a whole, and hence the lack of integration in a science concerned with this kind of whole, is merely evidence that its study is in an early stage. In the process of cognition of the whole, its parts will inevitably be brought out, and there will be a differentiation of the science, at which point it will inevitably assume an integrated character.

49. Influence on the life of human society, of course, does not come only from the natural elements of the earth's landscape envelope. Society is influenced even more by forces of nature whose sources are located outside of the landscape envelope. But these external influences (above all the sun) act on the life of society by means of the geographic environment and are for this reason often perceived as the effect of natural elements of the earth's landscape envelope.

50. This approach, which raises the operation of economic laws to an absolute, may be called *economic determinism* (economic materialism), which is the opposite of geographical determinism but is no less erroneous.

51. F. Engels, *Pismo I. Blokhu 21-22 sentyabrya 1890 g.* [Letter to J. Bloch of September 21-22, 1890], in K. Marx and F. Engels, *Izbranniye pisma* [Selected Letters] (Moscow, 1953), p. 422.

52. Ibid., p. 423.

53. Ibid., p. 424.

54. Most often the slowdown in the development of such countries is attributable to the influence of geographic conditions that isolate the country from the rest of the world. It is this kind of influence by the geographic factor that in large measure explains to us the situation in which even today one can find tribes in certain corners of the globe that are sometimes at the stage of the lowest or a middle level of barbarism or even savagery.

55. V. I. Lenin, *Detskaya bolezn "levizny" v kommunizme* [The Infantile Disease of Left-Wing Communism], in *Sochineniya*, 4th ed., vol. 31, p. 46.

56. This may be observed in geography with regard to the evaluation of local conditions. The same economic geographers who at one stage tried to prove that local conditions were not of substantial importance are now prepared to take into consideration nothing but local conditions.

57. K. Marx, *Ekonomichesko-filosofskiye rukopisi 1844 goda* [Economic-Philosophic Manuscripts from 1844], in K. Marx and F. Engels, *Iz rannikh proizvedeny* [From Early Works] (Moscow, 1956), p. 596.

58. For example, instead of being scientifically criticized and analyzed, the works of C. Ritter and A. Hettner are frequently subjected in our country to sheer vilification.

59. K. Marx, *Das Kapital*, vol. 1, p. 188.

60. Ibid., vol. 3, p. 833.

61. F. Engels, Pismo G. Shtarkenburgu, 25 yanvarya 1894 g. [Letter to H. Starkenburg, January 25, 1894], in K. Marx and F. Engels, *Izbranniye pisma*, p. 469.

62. Mao Tse-tung, *Otnositelno protivorechia* [Regarding Contradictions], in *Bolshevik*, 1952, no. 9, p. 11.

63. F. Engels, *Nachalo kontsa Avstrii* [The Beginning of the End for Austria], in K. Marx and F. Engels, *Sochineniya*, 2nd ed., vol. 4, p. 472.

64. K. Marx, *Ekonomichesko-filosofskiye rukopisi 1844 goda*, in K. Marx and F. Engels, *Iz rannikh proizvedeny*, p. 595.

65. K. Marx, *Das Kapital*, vol. 1, p. 381.

66. G. V. Plekhanov, *Osnovniye voprosy marksizma* [Basic Questions of Marxism], in *Izbranniye filosofskiye proizvedenia* [Selected Works of Philosophy], vol. 3 (Moscow, 1957), p. 153.

67. G. V. Plekhanov, *Istoria russkoi obshchestvennoi mysli* [The History of Russian Social Thought], in *Sochineniya*, vol. 20 (Moscow, 1925), p. 99.

68. G. V. Plekhanov, *K voprosu o razvitii monisticheskogo vzglyada na istoriyu* [On the Question of the Development of the Monistic View of History], in *Izbranniye filosofskiye proizvedenia*, vol. 1, p. 689.

CHAPTER SIX

1. Even paleogeography, which is concerned with the earth's landscape envelope of the distant past, does this in order to explain the causes of the origin and formation of the modern geographic environment. Thus in a certain (historical) sense paleogeography is concerned with the same geographic environment and hence has a certain practical value.

2. Many scholars quite rightly have called attention to this peculiarity of geography. See, for example, E. Martonne, *Osnovy fizicheskoi geografii* [Fundamentals of Phsyical Geography], vol. 1 (Moscow, 1939), pp. 26–30.

3. Of course, when geographical works are reduced to the mere description of the landscape envelope and its elements, this gives science much less than a description accompanied by profound analysis or by synthesis. But it would be totally wrong to contrapose descriptive and scientific geography. Descriptive geography without scientific value does not exist in real life and never will. Description is completely necessary in any science, and all the more so in geography.

4. Zdeněk Nejedlý, *Istoria cheshskogo naroda* [History of the Czech People], vol. 1, Moscow, 1952.

5. This becomes especially necessary when historians investigating social development shift from the general to the particular and when history ascertains, for example, the reasons for the uneven development of certain countries and regions.

6. It should be repeated that the geographical method is a narrower concept than the spatial method. Geography deals only with the surface of the earth. The landscape envelopes of other planets must be studied by the special science of astrography, which undoubtedly has a great future. Such sciences as selenography, marsography, and so forth are likely to begin to develop very soon.

7. S. V. Kalesnik, *Osnovy obshchevo zemlevedenia* [Fundamentals of General Earth Science] (Moscow, 1955).

8. M. V. Lomonosov, *Izbranniye filosofskiye proizvedenia* [Selected Works of Philosophy] (Moscow, 1950), p. 396.

9. C. Ritter, *Ob istoricheskom elemente v nauke zemlevedenia* [On the Historical Element in Earth Science], in *Magazin zemlevedenia i puteshestvy* [Jouranl of Earth Science and Travels], vol. 2, p. 482.

10. Chorology, territoriality, location, and other concepts of this type inevitably lead to unscientific conclusions when they are considered the *subject matter* of geography. Specialized chorological, territorial, and locational sciences do not exist and cannot exist.

But these same concepts are quite appropriate if they are viewed as basic methodological categories of geography. With reference to the methodological nature and not the subject-matter nature of the geographical sciences, one can safely say that without chorology (and, correspondingly, without territoriality and location) geographical research is impossible, since these concepts constitute the geographical method. And only the combination of a subject matter with a definite method defines the specificity of each science. In this sense it is completely valid to say that without chorology there is no geography.

11. F. Engels, *Anti-Dyuring* [Anti-Dühring], pp. 20–21.

12. The delineation of objective physical-geographic and especially economic-geographic regions is highly complex. It is easy here to make subjective errors. It should also be kept in mind that economic-geographic regions are often called economic regions in our literature.

13. Of the works by Soviet geographers dealing with the problem of economic geographic regionalization, the writings of N. N. Kolosovsky are of especially great importance.

14. In this case we mean the major or, in other words, basic regions, whose boundaries are defined in our country by the USSR State Planning Committee.

15. It should be taken into consideration that neither a commonly accepted terminology nor a taxonomic system of regionalization has been developed yet in economic geography. The word *region* is used in the most diversified senses in economic geography. There is no taxonomic order in state planning yet, either. In one case region is understood to mean enormous territories, in another, small territorial units, all the way down to cities. Administrative-economic territorial units or administrative-economic regions in the system of state economic regionalization are obviously units of a second order, that is, they are subregions with respect to the major economic-geographic regions. For a successful effort to create a taxonomic system of regionalization, see A. M. Kolotiyevsky, *O taksonomii, vidakh i raznovidnostyakh ekonomicheskogo raionirovania SSSR* [On the Taxonomy, Types and Varieties of Economic Regionalization of the USSR], in *Uchoniye zapiski Latvyskovo gosudarstvennovo universiteta im. P. Stuchki* [Latvian State University Scholarly Notes], vol. 27, no. 1 (1959), pp. 5–19.

16. In an actual setting such economic-geographic microregions take shape most often under conditions of a definite physical-geographic unity: in a mountain basin, where conditions of natural seclusion cause the formation of an area with specific economic features; in a desert oasis; around a lake; on a section of a river's flood plain; in a naturally bounded foothill zone; and so forth.

17. Yu. G. Saushkin, *Rabota Karla Marksa nad trudami russkogo geografa P. P. Semenova-Tyan-Shanskovo* [Karl Marx's Work on the Writings of the Russian Geographer P. P. Semenov-Tyan-Shansky], in *Voprosy geografii* [Questions of Geography], no. 31 (Moscow, 1953), pp. 123–31.

18. *KPSS v resolutsiakh i resheniakh syezdov, konferentsy i plenumov TsK* [The CPSU in Resolutions and Decisions from Congresses, Conferences, and Plenary Sessions of the Central Committee], part 1 (Moscow, 1954), pp. 718–19.

19. Ibid., part 2 (Moscow, 1954), p. 465.

20. Ibid., part 3 (Moscow, 1955), pp. 354–55.

21. The Directives of the Twentieth Congress for the Fifth Five-Year Plan for the Development of the USSR National Economy say: "To ensure improvement of the geographic location of the construction of industrial enterprises in the new five-year plan, with the purpose of bringing industry closer to the sources of raw materials and fuel so as to eliminate irrational and excessively long shipments." Ibid., p. 556.

22. N. S. Khrushchev, "O kontrolnykh tsifrakh razvitia narodnogo khozyaistva SSSR na 1959–1965 gody. Vneocherednoi XXI syezd Kommunist-icheskoi partii Sovetskogo Soyuza. Stenografichesky otchot." [On the Planned Figures for the Development of the USSR National Economy in 1959–1965. Extraordinary Twenty-First Congress of the Communist Party in the Soviet Union. Stenographic Transcript], part 1 (Moscow, 1959), p. 44.

23. N. N. Kolosovsky, *Osnovy ekonomicheskovo raionirovania* [Foundations of Economic Regionalization] (Moscow, 1958), p. 95.

24. Ibid., p. 39.

25. K. Marx and F. Engels, *Nemetskaya ideologia* [German Ideology], in *Sochineniya* [Works], 2d ed., vol. 3, p. 31.

26. V. I. Lenin, *Razvitiye kapitalizma v Rossii* [The Development of Capitalism in Russia], in *Sochineniya*, 5th ed., p. 53.

27. A. M. Rumyantsev, *Predmet politicheskoi ekonomii i kharakter zakonov ekonomicheskogo razvitia obshchestva* [The Subject Matter of Political Economy and the Character of the Laws of Economic Development of Society], in *Voprosy filosofii* [Questions of Philosophy], 1955, no. 2, p. 93.

28. In the process of production there is unquestionably a link (or, rather, an interpenetration) between the geographical, economic, and technical sciences. The geographical sciences are concerned with the conditions of production, the geographic environment in which the process of production takes place and which concurrently is material for this production. The economic sciences are concerned with intrasocietal relations in terms of production. The technical sciences are concerned with the tools of production. But this link between three different systems of sciences, though it underscores the unity of science as a whole and relates them to each other, does not rule out fundamentally important differences between them, defined above all by the difference in subject matter.

29. Although it makes a specialized study of production relations, political economy cannot, of course, ignore productive forces, which are inseparable from production relations. The same holds in economic geography, which cannot study territorial complexes of productive forces (in combination with instruments of labor) completely without reference to production relations. Not only does this proposition not contradict the substantive differences between political economy and economic geography, it affords a real possibility of cognizing their specific subject matter. Inter-penetration between sciences is not equivalent to their merger.

30. See, for example, M. S. Buyanovsky, *K voprosu o metodologicheskikh os-novakh ekonomicheskoi geografii* [On the Question of the Methodological Foundations of Economic Geography], in *Izvestia VGO* [News of the All-Union Geographical Society], 1954, no. 6, pp. 526–31.

31. Countries with the same social system can have (and usually do have) significant differences both in the level of development and in the structure of productive forces. These differences evolve under the influence of fundamentally diverse factors, of which the most important are: the degree of development of the social system and the level of development of technology; the specific character of natural resources (both in quantity and in quality); the specific character of natural conditions;

the characteristics of the cultures of peoples living in given countries. All of the above factors that determine differences in the productive forces of countries with the same social system are the result of distinctive features in the process of the historical development of nature and society, peculiarities that are inevitable concomitants of every country and every people.

32. Walter Scott's novels are obscure to Chinese readers, since feudalism in China took a completely different form than in England (no castles, jousting tournaments, etc.).

33. G. V. Plekhanov, *Istoria russkoi obshchestvennoi mysli* [The History of Russian Social Thought], in *Sochineniya*, vol. 20, p. 14.

34. Yu. G. Saushkin, *Vvedeniye v ekonomicheskuyu geografiyu* [Introduction to Economic Geography] (Moscow, 1958), p. 32.

35. D. S. Timoshkin, M. N. Khromov, P. T. Tikhonov, and M. A. Izrailev, *O predmete i zadachakh ekonomicheskoi geografii* [On the Subject Matter and the Tasks of Economic Geography], in *Izvestia VGO*, 1954, no. 5, p. 436.

36. Ibid., p. 438.

37. *K itogam diskussii po voprosam fizicheskoi i ekonomicheskoi geografii* [On the Results of the Debate on Questions of Physical and Economic Geography], in *Voprosy filosofii*, 1954, no. 5, p. 169.

38. V. M. Volpe and V. S. Klupt, *Lektsii po ekonomicheskoi geografii SSSR* [Lectures on the Economic Geography of the USSR], part 1 (Leningrad, 1957), p. 3.

39. Ya. G. Feigin, *O predmete i zadachakh ekonomicheskoi geografii* [On the Subject Matter and the Tasks of Economic Geography], in *Voprosy filosofii*, 1951, no. 6, p. 150.

40. N. Ya. Seversky, *Po voprosu o "yedinoi geografii"* [On the Question of a "Unified Geography"], in *Voprosy filosofii*, 1952, no. 2, p. 239.

41. An example is M. I. Albrut and A. P. Stukov, *Protiv teoreticheskoi putanitsy i propagandy "yedinoi geografii"* [Against the Theoretical Confusion and Propaganda of a "Unified Geography"] in *Izvestia AN SSSR* [News of the USSR Academy of Sciences], geographical series, 1959, no. 6, pp. 122–27.

42. See, for example, V. S. Molodtsov, *Ob oshibkakh v ponimanii predmeta dialekticheskogo materializma* [On Errors in the Understanding of the Subject Matter of Dialectical Materialism], in *Voprosy filosofii*, 1956, no. 1, pp. 188–94.

43. F. Engels, *Dialektika prirody* [The Dialectics of Nature], p. 141.

44. Physical geographers define the subject matter of physical geography in different ways. The more widespread definitions are the following: the geographic envelope, the landscape envelope, and the geographic environment. Despite the differences between these definitions, they all attest to a concrete, objectively existing subject matter.

45. V. F. Vasyutin's definition, in *Voprosy ekonomiki* [Problems of Economics], 1948, no. 8, p. 104.

46. A. D. Danilov's definition, quoted from V. A. Vityazev and V. S. Preobrazhensky, *O voprosakh geograficheskoi nauki* [On Questions of Geography], in *Voprosy filosofii*, 1951, no. 3, p. 174.

47. M. S. Buyanovksy, *K voprosu o metodologicheskikh osnovakh ekonomicheskoi geografii* [On the Question of the Methodological Foundation of Economic Geography], in *Izvestia VGO*, 1954, no. 6, p. 526.

48. Decision of the Second Congress of the USSR Geographical Society, in *Izvestia VGO*, 1955, no. 2, p. 97.

49. Report on the work of the Second Congress of the USSR Geographical Society, in *Izvestia VGO*, 1955, no. 2, pp. 93–95.

50. It should not be inferred from this that geographers should not deal with questions of location of production at all. They should do so together with representatives of other sciences.

51. N. N. Baranskiy, *Uchot prirodnoi sredy v ekonomicheskoi geografii* [Taking Account of the Natural Environment in Economic Geography], in *Ekonomicheskaya geografia. Ekonomicheskaya kartografia* [Economic Geography and Economic Cartography], pp. 41–42.

52. I. M. Zabelin recently made an attempt at a theoretical substantiation of this kind of split. See his book, *Teoria fizicheskoi geografii* [The Theory of Physical Geography] (Moscow, 1959).

53. V. E. Den, *Kurs ekonomicheskoi geografii* [Course in Economic Geography] (Leningrad and Moscow, 1925), pp. 3–4.

54. N. N. Kolosovsky, *Nauchniye problemy geografii* [Scientific Problems of Geography], in *Voprosy geografii* [Questions of Geography], no. 37, p. 144.

55. O. A. Konstantinov, *K istorii i sovremennomu sostoyaniyu ekonomiko-geograficheskoi nauki SSSR* [On the History and the Present State of Economic Geography in the USSR], in *Izvestia VGO*, 1955, no. 3, p. 266.

56. *Vporosy geografii,* no. 37 (Moscow, 1955).

57. I. A. Vitver, *Dezorganizuyushchiye retsenzii* [Disorganizing Reviews], in *Geografia v shkole* [Geography in School], 1935, no. 4, pp. 66–76.

CHAPTER SEVEN

1. Hegel, *Nauka logiki* [The Science of Logic], in *Sochineniya* [Works], vol. 1, p. 81.

2. D. V. Nalivkin, *I. V. Mushketov i geograficheskaya ideologia* [I. V. Mushketov and Geographical Ideology], in *Izvestia VGO* [News of the All-Union Geographical Society], 1952, no. 3, p. 244.

3. F. Engels, *Anti-Dyuring* [Anti-Dühring], p. 40.

4. The question sometimes arises of whether geographers should study differences within individual enterprises. It seems to us that the answer to this question should depend on the character of the enterprise. If the location of certain parts of the enterprise is not related to differences in the geographic environment, then that is technical-economic location, which geography is not concerned with (e.g., the location of shops in a factory). But when an enterprise has units that are located in different geographic conditions and that have their own specific characteristics in relation to the environment, these units of an enterprise can and should be objects of geographical study (e.g., the location of production on a collective or state farm, the location of individual units of large-scale hydraulic installations, and so forth).

5. The division of geography not only into subfields but also into parts has been known for a very long time. The essence of this kind of division was shown very clearly by Bernard Varenius. However, it is still necessary to mention this now, since the division of geography into parts is sometimes completely overlooked because of the overly intensified differentiation of science.

6. In comparing the two editions of S. V. Kalesnik's work, one cannot fail to note that the second edition is an unmistakable step in the direction of general earth science. In our view, the inclusion of man in earth science made S. V. Kalesnik's book a notable contribution to geography. The value of S. V. Kalesnik's work, of course, goes far beyond academic significance.

7. It would probably be most appropriate for the majority of states to propose a trilevel taxonomic system: state–country(*strana*)–region. But perhaps it would be

more suitable for the USSR, the Chinese People's Republic, and maybe the other larger states to use a four-level division, to wit: (1) a union of countries; (2) a union country; (3) country; (4) region. (Example: USSR, the Ukraine Republic, the Western Ukraine, Transcarpathian Province.)

8. J. V. Stalin, *Marksizm i natsionalny vopros* [Marxism and the Nationalities Question], in *Sochineniya*, vol. 2, p. 296.

9. But this influence was not direct, contrary to the picture that advocates of geographical determinism have drawn.

10. N. N. Baranskiy, *Stranovedeniye i geografia fizicheskaya i ekonomicheskaya* [Regional Geography and Physical and Economic Geography], in *Ekonomicheskaya geografia. Ekonomicheskaya kartografia* [Economic Geography and Economic Cartography], p. 165.

11. Ibid., p. 157.

12. The definition of regional geography as a mere "organizational form" is clearly at variance with everything else in the cited article by N. N. Baranskiy. In fact, it is even more at variance with all of his other statements about the unity of geography, which were directed against mechanical unifications of different sciences and against mechanical approaches in all their manifestations. It is characteristic of the opponents of regional geography that they very often make use of this incorrect formulation by N. N. Baranskiy against the basic trend of his works.

13. V. I. Lenin, *K voprosu o dialektike* [On the Question of Dialectics], in *Sochineniya* [Works], 4th ed., vol. 38, p. 361.

14. Instead of geographical educational and research institutions, various nongeographical institutions (institutes for Eastern studies, regional geography departments in some linguistic academic institutions, the Institute of International Relations and others) are attempting to meet the practical need for regional specialists. Although such nongeographical institutions are necessary, it must be said that they cannot train specialists with an all-around knowledge of foreign countries. To improve this situation, a specialized regional section should be organized in some geography faculties at universities (e.g., at Moscow, Leningrad and Kiev Universities) and at least one specialized regional (in the geographical sense) research institute should be created in the system of the USSR Academy of Sciences.

15. Ya. G. Feigin, *O predmete i zadachakh ekonomicheskoi geografii* [On the Subject Matter and the Tasks of Economic Geography], in *Voprosy filosofii* [Questions of Philosophy], 1951, no. 6, pp. 150–59.

16. V. I. Lenin, *K voprosu o dialektike*, in *Sochineniya*, 4th ed., vol. 38, p. 359.

17. V. I. Lenin, *Razvitiye kapitalizma v Rossii* [The Development of Capitalism in Russia], in *Sochineniya*, 5th ed., vol. 3, pp. 308–9.

18. I. M. Zabelin, *Teoria fizicheskoi geografii* [A Theory of Physical Geography] (Moscow, 1959), p. 280.

19. Ibid., p. 281.

20. Mao Tse-tung, *Otnositelno praktiki* [Regarding Practice], in *Izbranniye proizvedenia* [Selected Works], vol. 1 (Moscow, 1952), p. 528.

21. N. S. Khrushchev, *O dalneishem uvelichenii proizvodstva zerna v strane i ob osvoyenii tselinnykh i zalezhnykh zemel* [On a Futher Increase in Grain Production in the Country and on Development of Virgin and Unused Lands] (Moscow, 1954), p. 18.

22. Yu. G. Simonov, *O primenenii metodov fizicheskoi geografii v stroitelstve* [On the Use of the Methods of Physical Geography in Construction], in *Geografia i khozyaistvo* [Geography and Economy], 1958, nos. 3–4, p. 28.

23. Ibid., p. 31.

24. K. V. Zvorykin and K. I. Ivanov, *Zadachi geograficheskogo izuchenia prirodnykh i ekonomicheskikh uslovy selskokhozyaistvennogo proizvodstva* [The Tasks of the Geographical Study of the Natural and Economic Conditions of Agricultural Production], in *Geografia i khozyaistvo,* 1958, no. 1, p. 37.

25. Ibid.

26. One of the examples that show how essential integrated geographical inquiry is in each instance of planned large-scale hydroelectric construction are the conclusions that the construction of large-scale hydroelectric centers in the lower reaches of the Ob and Irtysh rivers would be unprofitable (see V. G. Konovalenko, *K voprosu gidroenergeticheskogo stroitelstva v nizhnem techenii reki Obi* [On the Question of Hydroelectric Construction in the Lower Reaches of the Ob River], in *Geografia i khozyaistvo,* 1958, nos. 3-4, pp. 38-41). Also see: G. S. Vyzgo, *O generalnom plane ispolzovania vodnykh, zemelnykh i energeticheskikh resursov* [On the General Plan for the Utilization of Water, Land and Electric-Power Resources], in *Geografia i khozyaistvo,* 1960, no. 6, pp. 3-13.

27. N. N. Kolosovsky, *Nauchniye problemy geografii* [Scientific Problems of Geography], in *Voprosy geografii* [Questions of Geography], no. 37, p. 144.

28. S. G. Kolesnev, *Organizatsia sotsialisticheskikh selskokhozyaistvennykh predpriyaty* [The Organization of Socialist Agricultural Enterprises] (Moscow, 1947), p. 11.

29. V. S. Nemchinov, *O kriteriakh razmeshchenia kultur i otraslei zhivotnovodstva* [On the Criteria for the Location of Crops and Livestock-Raising Sectors], in *Izvestia AN SSSR* [News of the USSR Academy of Sciences], Economics and Law Section, 1947, no. 2, p. 76.

30. V. F. Vasyutin (speech at session of Academic Council of the USSR Academy of Sciences Institute of Economics), in *Voprosy ekonomiki* [Problems of Economics], 1948, no. 8, p. 108.

31. Ibid., p. 109.

32. Ibid.

33. *Pravda,* February 17, 1960.

34. A. I Tulupnikov, *Ob ekonomicheskom obrazovanii sistem zemledelia i zhivotnovodstva pri razrabotke ratsionalnykh sistem vedenia selskovo khozyaistva* [On the Economic Formation of the Farming and Livestock-Raising Systems in the Development of Rational Systems for the Management of Agriculture], in *Doklady i soobshchenia Vsesoyuznogo nauchno-issledovatelskogo instituta ekonomiki selskogo khozyaistva* [Reports and Accounts from the All-Union Research Institute of Agricultural Economics], no. 3 (Moscow, 1958), p. 17.

35. "From the Editorial Board," in *Geografia i khozyaistvo,* 1958, no. 1, p. 3.

36. F. Engels, *Anti-Dyuring* [Anti-Dühring], p. 280.

37. For example, the system of agricultural management was completely without geographers capable of doing an economic land appraisal. Economic land appraisals are still done in our country case by case. As yet there is no special geographical organization to deal systematically with this important practical-scientific work.

38. The specializations of economic geography and physical geography exist in all university geography faculties, but regional specialization is extremely inadequate — the students are trained not as integrated geographers but as geography teachers, that is, the function of pedagogical institutes is duplicated.

39. A. M. Ryabchikov, *O putyakh razvitia universitetskoi geografii* [On the Paths of Development of University Geography], in *Vestnik MGU* [Moscow University News], series 5, 1960, no. 1, p. 6.

40. The term (concept) *local studies* [*krayevedeniye*] is also used in a different

sense. It signifies sets of highly diversified information about individual localities. We believe the local-studies movement would gain much if it would take as its basis the geographical study of small territories, which would rid it of aimless universalism and would provide an opportunity to direct it toward servicing economic practice. It would also be advisable, in our view, to reorganize museums of local studies into museums of the history and geography of their respective territories or provinces. For example, Ryazan should have a "Ryazan Province Geographical-Historical Museum." What is involved here, of course, is not a mere change of signs, but a change in the substance of the work of these museums, which should be research institutions that also resolve practical scientific tasks most relevant to the province and not be simple collections of more or less well-selected exhibits.

41. P. N. Stepanov, *Razmeshcheniye proizvoditelnykh sil i krayevedeniye* [The Distribution of Productive Forces and Local Studies], in *Sovetskoye krayevedeniye* [Soviet Local Studies], 1932, no. 6, p. 5.

42. The specialized bias in regional works must not be carried to an extreme, to a loss of integration. Regional studies cannot be departmental in the character of its subject matter (countries, regions) or in the method it employs. Although they pay more attention to individual aspects of the geographic environment, regional works should shed light on specific questions (transport, agricultural, etc.) in their interconnection with all the other geographic conditions that have evolved on a given territory.

43. For example, regional works with the most successful synthesis of concrete material on the geographic environment of certain countries include the following: E. M. Murzayev, *Mongolskaya Narodnaya Respublika* [The Mongolian People's Republic] (Moscow, 1952); S. N. Ryazantsev, *Kirgizia* (Moscow, 1951); V. T. Zaichikov, *Koreya* [Korea] (Moscow, 1951); E. B. Valev, *Bolgaria* (Moscow, 1957).

44. "From the Editorial Board," *Geografia i khozyaistvo,* 1958, no. 1, p. 3.

PART THREE
APPENDIXES

APPENDIX ONE

Summary of Major Conclusions

The major conclusions presented by the author [taken from Chapters 5, 6, and 7, Eds.] are listed below for the convenience of the reader.

1. The object of study of all the geographical sciences is a concrete form of the material world. The subject matter of geography is the landscape (or geographic) envelope (sphere) of the earth.

2. The part of the earth's landscape envelope *within* which human society originated and develops is called the *geographic environment.*

At present there is practically no essential difference between the geographic environment and the landscape envelope, within which the life of human society takes place. Therefore, with the exception of certain specialized divisions, all the geographical sciences, like geography as a whole, have their common object of inquiry precisely in the geographic environment.

3. The geographic environment consists of three groups (complexes) of elements. The development of each of them is governed by their own specific conformity with law. The first group (inorganic) develops under the influence of physical-chemical laws. The development of the second group (organic) is governed by biological laws. The third group (social) develops under the determining influence of social laws.

The geographic environment is a complex, contradictory unity in which a struggle of opposites takes place. This struggle, primarily between complexes of its elements, is the chief force determining the internal causes of the development of the geographic environment as a whole.

4. The laws that govern the development of the inorganic complex of elements of the geographic environment continue to operate both in the group of organic and in the group of social elements. The laws that govern the development of the

organic group of elements continue to operate in human society (in the group of social elements).

This fundamentally important fact, which unifies all the elements of the geographic environment, is a most important basis for the constant links between all things and phenomena of the material world. It also allows one to find *properties common to all forms of development of matter.*

But these laws do not operate in reverse. Social laws do not operate in the biological sphere or in the inorganic complex of the geographic environment. Biological laws, in turn, do not operate in the inorganic complex. Thus, *the more complex and the higher the form of development of matter* in the geographic environment, the *more complex the combination of patterns influencing this development.* At the same time, the higher the form of motion of matter, the greater number of laws of development of the material world it *possesses for its reproduction,* subject to the determining influence of its specific patterns.

5. The complex, contradictory character of the geographic environment defines the complex (integrated) character of the science concerned with it. The group of inorganic elements of the geographic environment is studied by physical geography. The group of organic elements of the geographic environment is studied by biological geography (which is developing in our country within physical geography). The group of social elements of the geographic environment is studied by social (economic) geography.

The three complexes of elements of the geographic environment are not separated by an impenetrable partition, there are intersections between them; and all of them put together constitute a *unified whole.* Therefore, in addition to the separate study of each complex of elements in the geographic environment, a science is needed that, by generalizing the research of the three geographies, could comprehend the geographic environment as a whole. The geographic environment is not only the sum of its constituent elements. Hence rejection of geography as an integrated science inevitably *leads to the conclusion that the geographic environment is unknow-*

able. No matter how many new geographical sciences arise in this connection, geography will always be indispensable as a science of the geographic environment as a whole. *The differentiation of geography cannot lead to its elimination.*

By consolidating all the sciences concerned with the elements of the geographic environment, geography is simultaneously a system of these sciences and a synthetic, integrated science of the geographic environment as a whole. Although it is the object of study for all the geographical sciences, the geographic environment is also the subject matter of geography, which, by generalizing and synthesizing the results of research by its elements, creates unified theories of it.

This, in our view, is the *subject-matter essence of the unity of geography*.

6. Geography is a science concerned with a structurally complex subject matter. Its differentiation is therefore an inevitable and completely necessary process. Historically this process went from the general to the particular. Arising from observations of separate occurrences, geography originally was not a differentiated science. Later, as it developed, accumulated geographical knowledge, and improved analytically, two branches were distinguished in geography: One concerned with the natural complex of the geographic environment, the other with its social complex. In our country these branches have developed as physical and economic geography.

Subsequently physical geography, in turn, spawned branches concerned with individual elements of the natural complex of the geographic environment, which then became their subject matter. Thus physical geography, though remaining a branch of geography as a whole, was itself transformed into an integrated science and simultaneously into a system of sciences. The same is happening to economic geography, although its process of specialization is slower.

However, this process of differentiation of geography is only *one side of the unified process of its development* — the *analytic* side. There is another side — the *synthetic* side. It is manifested in the increasing interpenetration between the geographical sciences, in the establishment of general patterns of development of the elements of the geographic environment, and in the effort to create integrated pictures of its territorial complexes.

Such is the historical course of development of geography, conditioned in a law-governed way by the unity of its subject matter and method. It is also conditioned by the unity of science as a whole, which does not recognize internal, insurmountable barriers at all. The unity of science attests to the relativity of its divisions and to the relativity of the classification of individual sciences.

7. The history of geography clearly illustrates its unity. For a long time physical and economic geography developed as one general science. They have a common history and prehistory. In the process of their subsequent differentiation they were reshaped quite naturally into special sciences with their own specific tasks and methods. At the same time geography as a whole retained the common goals of inquiry that cannot bring its individual divisions and branches (physical and economic geography in particular) to a complete separation. Geography as an integrated science will inevitably continue to develop on the basis of the achievements of all its constituent branches.

The history of geography vividly contradicts the spokesmen for dualism in geography, the exponents of two geographies. It is precisely for this reason that they are forced to reject this history. They aver that geography before the mid-nineteenth century was not a science. The science, in their opinion, took shape only when the process of differentiation of geography occurred. Moreover, efforts are still being made in our country to deprive economic geography even of this abbreviated history.

This attitude toward the history of economic geography is a striking manifestation of nihilism, concealed most often by high-sounding phrases alleging that economic geography could not have existed before the appearance of Marxism. No one can deny that Marxism and the construction of socialism have added and continue to add new content to economic geography, and it is simply impossible on this basis to deny the existence of economic geography in the past (even in the fairly remote past).

Seeing only one side of the development of science, the opponents of integrated geography are playing the role of its liquidators, attempting thereby to deny the objective process of the development of science as a combination of analysis and synthesis.

Yet it is precisely now that geography is passing through a pivotal period, in which the predominant analytic trend in its development is no longer sustaining the process of cognition of the geographic environment. The enormous amount of empirical data that has accumulated at present is not receiving the necessary generalization because of the scarcity of synthetic research and the lack of a detailed theory of integrated geography as an integrated science. Therefore the denial of geography's unity is now especially deleterious not only to theory because by incorrectly orienting scholars, it also gives geographical research an unbalanced direction.

8. Familiarity with the history of geography shows that the theories of geographers were always closely associated with various philosophic concepts. The materialistic trend in philosophy was the basis of the deterministic world view in geography. The mechanistic character and inconsistency of pre-Marxian materialism were manifested in geography in the form of geographical determinism. Whereas Soviet geography adopts determinism as one of the most necessary aspects of dialectical materialism, it discards *geographical determinism,* which in our day has completely demonstrated its scientific untenability and is a basis for the development of various pseudoscientific theories.

The idealistic trend in philosophy was the basis for the indeterministic views of geographers. In every epoch indeterminism *impeded* the development of science, strengthened its *tendency toward exclusive empiricism,* and led to fragmentation, to the establishment of an artificial separation between sciences, and to the total opposition of social science to natural science. Indeterminism has always denied the possibility of a monistic world view and has denied the possibility of the development of geography as an integrated science with a common object of study and a single methodological basis.

9. The unity of geography is defined not only by a common object of study but also by a certain *unity in the method applied.* All of the geographical sciences, like geography as a whole, investigate their subject matter through its *territorial complexes.* The geographical sciences can therefore be called *sciences of territorial complexes. The combination of the geographical method with specific subject matters,* all of which are either elements or complexes of elements of the earth's landscape envelope, *consolidates* the geographical sciences into a single *system,* thereby distinguishing it from the other scientific systems related to geography (economic, technical, biological, etc.).

Herein lies, in our view, the *methodological essence of the unity of geography.*

A specific form of the geographical method is *regionalization,* without which geographical inquiry is inconceivable. Another specific form of the geographic method is *mapping,* which is also an indispensable attribute of any specific geographical inquiry.

10. In studying the social (economic) elements of the geographic environment and their territorial complexes, economic geography evinces a significant (substantive) difference from the economic sciences, which study production relations. For this reason economic geography cannot possibly be categorized as an economic science, which would remove it from the system of geographical sciences. A classification of sciences should be based on a *combination of the subject matter studied by the given science* and the *basic method* it employs.

11. An incorrect definition of the subject matter of economic geography (as location) not only leads to the separation of economic geography from the geographical sciences but also deprives it of a *material subject matter.* In essence, the locational definition of economic geography is a revival in somewhat altered form of the old chorological concepts that

put chorology, that is, specificity in the approach to a study, in the place of *subject matter*. In other words, the material subject matter is replaced by a methodological category.

The locational conception of economic geography develops into a rejection of the determinist view of the world. By raising the laws of social development to an absolute, it removes human society from the material world of nature. Therefore, the development of the locational definition of economic geography leads ultimately and inevitably to indeterminist conclusions.

12. The unity of geography's subject matter, in combination with the geographical method, makes it possible to inquire into the geographic environment not only by elements but also by their complexes, all the way up to the complete complex of all the elements in the geographic environment. This theoretically proves the possibility of autonomous integrated geographical research. Integrated studies of this kind, especially on small territories, can be highly diversified: they may be general geographical, physical-geographical, biogeographical, economic-geographical, and specialized (applied-science) works. The combination of the elements studied in territorial complexes is determined in each case by the *purpose of inquiry*.

13. Geography is a science of territorial complexes. Consequently, by its very nature it can be at different scales. The geographical method, depending on the scale of inquiry, allows for the determination of territorial complexes with the most diverse degrees of generalization. The degree of generalization is of exceptional importance in geography.

Whereas the complex structure of the geographic environment gives rise to the *branch* differentiation of geography, the degree of generalization (scale of inquiry) gives rise to *its division into parts*. At the present stage of scientific development it is most expedient, obviously, to divide geography into three parts: *earth science, regional geography,* and *local geography*. Each of these parts, though not forming a distinct science, has important features peculiar to it. Therefore it is expedient for geography, along with its branch specializations, to specialize on the basis of its parts. In other words, the vastness of concrete material and its complexity demand *concrete regional specialization* for the integrated geographic approach.

Herein lies one of the important distinctions of the broad geographical approach as opposed to the branch approach, which does not have such an acute need for regional specialization. It is relatively easy for geomorphologists, climatologists, hydrologists, and others, to change territory when they study their subject matter, which cannot be said about physical geographers, economic geographers, and especially regional geographers, since they deal not with individual elements but with complexes of elements of the geographic environment, which will have incomparably more differences from territory to territory than a member of any thematic geography will encounter. The complexes of the geographic environment, as a rule, are unique. Territories with very similar relief (or with a very similar complexes of the geographic environment are impossible to find.

14. Geography is the science of the territorial complexes of the geographic environment. By determining and studying these complexes, it shows ways to make fuller and more rational use of them in practice.

By its very nature, geography is supposed to equip practitioners with knowledge of geographic conditions, knowledge of the specificity of these conditions from place to place. This is the basic practical significance both of geography as a whole and of any individual geographical science as a whole.

15. The practical importance of geography is becoming especially significant in the epoch of socialism, when the practical application of the results of geographical research is not hampered by the private-ownership mode of production.

In the conditions of the socialist socioeconomic system, which is directed toward maximum utilization of natural and social conditions and resources in the interests of mankind, unprecedented opportunities open up for the further subordination of nature to the interests of society.

Planned management of the economy has become possible

under socialism, and because of this it has become especially important to have ample knowledge of the territorial complexes of the geographic environment. Without this knowledge, glaring errors in the long-range planning of the national economies of socialist countries are inevitable.

Taking account of territorial differences in the geographic environment rids applied work of stereotyped planning and promotes the correct integration of the branches of the national economy on a regional basis. However by taking account of the geographic environment implies knowing it, and practitioners can be equipped with knowledge of the territorial complexes of the geographic environment only by an integrated science — *geography.*

The underestimation of geography in practice, which reflects a failure to understand its unity, has done and will continue to do noticeable harm to the national economy, by retarding its development and *lowering the productivity of labor.*

16. The successful implementation of practical-scientific, integrated geographical research at the present stage of development of Soviet geography is connected in the closest way with the necessity of resolving a number of theoretical questions. The correct resolution of the theoretical question of the *unity* of geography is of especially great *practical* significance. This is based on the fact that the still widespread notion denying the unity of geography (and thereby denying the possibility of understanding the territorial complexes of the geographic environment) hampers the implementation of integrated, especially general, geographical research. The denial of the unity of geography theoretically justifies the exclusively thematic tendency in its development, and the elevation to an absolute of the general laws of development leads to a disregard of local peculiarities in natural and social conditions and to a stereotyped approach in economic practice that the Communist Party has denounced.

17. Elaboration of a theory of geography on the basis of Marxist-Leninist philosophy should be combined with more intensive development of integrated geographical works,

above all those of an applied-science character.

In the process, the following trends should assume leading significance:

a. comprehensive study of the geographic environment of small areas for the purpose of satisfying the immediate needs of practice (the agricultural and silvicultural appraisal of land, the study of water resources, the integrated study of regions with construction on the largest scale, etc.);

b. organization of effective control over correct utilization of the geographic environment in the process of production. Prevention of the squandering of natural resources;

c. determination and study of large economic-geographic regions. Elaboration of their characterizations, so that they may serve as an actual basis for long-range planning of the development of the economies of the USSR and the Union republics;

d. determination and characterization of economic-geographic regions (of the second order) with appraisal of their geographic conditions for the needs of regional economic councils;

e. development of integrated expeditions for the study of the deleterious consequences of irrational economic use of the geographic environment (erosion and so forth) with elaboration of a system of measures to reconstruct, preserve, and enrich it;

f. codified geographical monographs in regional geography, and above all a "Grand Geography of the USSR," and regional characterizations of foreign countries;

g. creation of public local-geographical mass organizations with the task of uncovering local resources and conditions for the needs of economic practice and the conservation of nature.

The above trends in the work of Soviet geographers, of course, are far from encompassing all the aspects of their study of the geographic environment. However these are the ones, in our view, that should be advanced at present as the principal ones, and special attention should be paid to them.

APPENDIX TWO

Selected Bibliography of Publications Relating to *Theoretical Problems of Geography* and the Ensuing Controversy That Have Appeared in English

1961

Al'brut, M. I "Let Us Clear Up for Once and for All Differences on Methodological Questions in Economic Geography." *Soviet Geography: Review and Translation,* vol. 2, no. 3 (March, 1961), pp. 23–25.

Anuchin, V. A. "On the Subject of Economic Geography." *Soviet Geography: Review and Translation,* vol. 2, no. 3 (March, 1961), pp. 26–34.

Baranskiy, N. N. "Book Review: V. A. Anuchin. *Teoreticheskiye problemy geografii* [Theoretical Problems of Geography]," vol. 2, no. 8 (October, 1961), pp. 81–84.

Konstantinov, O. A. "A Methodological Jumble in Theoretical Problems of Geography." *Soviet Geography: Review and Translation,* vol. 2, no. 10 (December, 1961), pp. 12–18.

Matley, I. "The Soviet Approach to Geography." Ph. D. dissertation, University of Michigan, 1961.

"Report on Discussion of Anuchin's Book." *Soviet Geography: Review and Translation,* vol. 2, no. 10 (December, 1961), pp. 18–32.

1962

Anuchin, V. A. "On the Criticism of the Unity of Geography." *Soviet Geography: Review and Translation,* vol. 3, no. 7 (September, 1962), pp. 22–39.

Furman, A. Ye. "On Interrelationships Between Natural and Social Laws." *Soviet Geography: Review and Translation,* vol. 3, no. 7 (September, 1962), pp. 49–54.

Hooson, D. J. M. "Methodological Clashes in Moscow." *Annals of the Association of American Geographers,* vol. 52, no. 4 (December, 1962), pp 469–75.

Kalesnik, S. V. "About 'Monism' and 'Dualism' in Soviet Geography." *Soviet Geography: Review and Translation,* vol. 3, no. 7 (September, 1962), pp. 3–16.

Kolosovskiy, N. N. "On the Concept of the Unity of Geography." *Soviet Geography: Review and Translation,* vol. 3, no. 7 (September, 1962), pp. 39–44.

Konovalenko, V. G. "The Concept of the Unity of Geography in the Solution of Basic Problems of Geography." *Soviet Geography: Review and Translation,* vol. 3, no. 7 (September, 1962), pp. 45–49.

Saushkin, Yu. G. "Economic Geography in the U.S.S.R." *Economic Geography,* vol. 38, no. 1 (January 1962), pp. 28–37.

Zakharov, N. D. "Scholastics Instead of Science." *Soviet Geography: Review and Translation,* vol. 3, no. 7 (September, 1962), pp. 16–22.

1963

Anuchin, V. A. "A New Book with Old Ideas." *Soviet Geography: Review and Translation,* vol. 4, no. 10 (December, 1963), pp. 71–72.

Alampiyev, P. M. and Ya G. Feygin, eds. "Methodological Problems in Economic Geography." *Soviet Geography: Review and Translation,* vol. 4, no. 10 (December, 1963), pp. 34–70.

"Author's Rebuttal of Saushkin's Reply." *Soviet Geography: Review and Translation,* vol. 4, no. 8 (October, 1963), pp. 31–34.

Avsyukh, G. A.; M. I. Budyko; I. P. Gerasimov; A. A. Grigor'yev; F. F. Davitaya; F. V. Kalesnik; and V. G. Sochava. "Geography in the System of Earth Sciences." *Soviet Geography: Review and Translation,* vol. 4, no. 8 (October, 1963), pp. 3–13.

Gladkov, N. A. "Defense of V. A. Anuchin's Doctoral Dissertation." *Soviet Geography: Review and Translation,* vol. 4, no. 8 (October, 1963), pp. 34–44.

Konovalenko, V. G. "About S. V. Kalesnik's Article on 'Monism' and 'Dualism' in Soviet Geography." *Soviet Geography: Review and Translation,* vol. 4, no. 10 (December, 1963), pp. 19–34.

"Letter of Protest Regarding Yu. G. Saushkin's Article in *Economic Geography.*" *Soviet Geography: Review and Translation,* vol. 4, no. 1 (January, 1963), pp. 60–62.

Pokshishevskiy, V. V. "On the Character of Laws in Economic Geography." *Soviet Geography: Review and Translation,* vol. 4, no. 4 (April, 1963), pp. 3–16.

Saushkin, Yu. G. "Reply to a Letter of Protest from a Soviet Geographer." *Soviet Geography: Review and Translation,* vol. 4, no. 8 (October, 1963), pp. 25-30.

_____ . "The Geographical Environment of Human Society." *Soviet Geography: Review and Translation,* vol. 4, no. 10 (December, 1963), pp. 3-19.

_____ . "V. A. Anuchin's Doctoral Dissertation Defense." *Soviet Geography: Review and Translation,* vol. 4, no. 1 (January, 1963), pp. 53-59.

Spate, O. H. K. "Theory and Practice in Soviet Geography." *Australian Geographical Studies,* vol. 1, no. 1 (April, 1963), pp. 18-30.

Vol'skiy, V. V. "Some of the Theories and Practice in Economic Geography." *Soviet Geography: Review and Translation,* vol. 4, no. 8 (October, 1963), pp. 14-25.

1964

Anuchin, V. A. "The Problem of Synthesis in Geographic Science." *Soviet Geography: Review and Translation,* vol. 5, no. 4 (April, 1964), pp. 34-46.

Konstantinov, F. V. "Interaction Between Nature and Society and Modern Geography." *Soviet Geography: Review and Translation,* vol. 5, no. 10 (December, 1964), pp. 61-73.

"L. F. Ilyichev's Remarks About a Unified Geography." *Soviet Geography: Review and Translation,* vol. 5, no. 4 (April, 1964), pp. 32-34.

Ryabchikov, A. M. "On the Interaction of the Geographical Sciences." *Soviet Geography: Review and Translation,* vol. 5, no. 10 (December, 1964), pp. 61-73.

Saushkin, Yu. G. "Methodological Problems in Soviet Geography as Interpreted by Some Foreign Geographers." *Soviet Geography: Review and Translation,* vol. 5, no. 8 (October, 1964), pp. 50-65.

Zhirmunskiy, M. M., and N. F. Yanitskiy. " 'Methodological Clashes in Moscow' as Misinterpreted by an American Geographer." *Soviet Geography: Review and Translation,* vol. 5, no. 6 (June, 1964), pp. 85-91.

1965

Anuchin, V. A. "A Sad Tale About Geography." *Soviet Geography: Review and Translation,* vol. 6, no. 7 (September, 1965), pp. 27-31.

Armand, D. L. "Let's Not." *Soviet Geography: Review and Translation,* vol. 6, no. 7 (September, 1965), pp. 32-36.

Chappell, J. E., Jr. "Marxism and Geography." *Problems of Communism,* vol. 14, no. 6 (November–December, 1965), pp. 12–22.

Demko, G. J. "Trends and Controversies in Soviet Geography." In *The State of Soviet Science,* edited by Walter Z. Laqueur, pp. 165–73. M.I.T. Press, Cambridge, 1965.

Gerasimov, I. P. "Has Geography 'Disappeared'?" *Soviet Geography: Review and Translation,* vol. 6, no. 7 (September, 1965), pp. 38–41.

Gokhman, V. M.; M. B. Gornung; and V. P. Kovalevskiy. "Not for the Sake of the Honor of the Uniform." *Soviet Geography: Review and Translation,* vol. 6, no. 7 (September, 1965), pp. 47–50.

Kalesnik, S. V. "Some Results of the New Discussion about a 'Unified' Geography." *Soviet Geography: Review and Translation,* vol. 6, no. 7 (September, 1965), pp. 11–26.

Saushkin, Yu. G. "The Today and Tomorrow of Geography." *Soviet Geography: Review and Translation,* vol. 6, no. 7 (September, 1965), pp. 50–56.

Zabelin, I. "What is the Matter with Geography?" *Soviet Geography: Review and Translation,* vol. 6, no. 7 (September, 1965), pp. 47–50.

1966

"Concerning the Discussion of Geography in *Literaturnaya Gazeta.*" *Soviet Geography: Review and Translation,* vol. 7, no. 2 (February, 1966), pp. 3–9.

Saushkin, Yu. G. "A History of Soviet Economic Geography." *Soviet Geography: Review and Translation,* vol. 7, no. 8 (October, 1966), pp. 3–98. See especially pages 82–86.

Zhirmunskiy, M. M. "The Interaction Between Nature and Society and Economic Geography." *Soviet Geography: Review and Translation,* vol. 7, no 7 (September, 1966), pp. 19–28.

1968

Gerasimov, I. P. "Fifty Years of Development of Soviet Geographic Thought." *Soviet Geography: Review and Translation,* vol. 9, no. 4 (April, 1968), pp. 238–52. See especially pages 247–48.

Gumilëv, L. N. "On the Subject of the 'Unified' Geography." *Soviet Geography: Review and Translation,* vol. 9, no. 1 (January, 1968), pp. 36–47.

1970

Anuchin, V. A. "Mathematization and the Geographic Method." *Soviet Geography: Review and Translation,* vol. 11, no. 2 (February, 1970), pp. 71–81.

———— . "On the Problems of Geography and the Task of Popularizing Geographical Knowledge." *Soviet Geography: Review and Translation,* vol. 11, no. 2 (February, 1970), pp. 82-112.

Semevskiy, B. N. "Propaganda for a Unified Geography." *Soviet Geography: Review and Translation,* vol. 11, no. 6 (June, 1970), pp. 501–8.

Yefremov, Yu. K. "The Scientific Basis for Presentation of the General Geographical Relations of the Landscape Sphere in an Exhibit of the Museum of Earth Science." *Soviet Geography: Review and Translation,* vol. 11, no. 2 (February, 1970), pp. 113–27.

1972

Anuchin, V. A. "Straddling the Boundaries Between Sciences." *Soviet Geography: Review and Translation,* vol. 13, no. 7 (September, 1972), pp. 432–40.

Isachenko, A. G. "On the Unity of Geography." *Soviet Geography: Review and Translation,* vol. 13, no. 4 (April, 1972), pp. 195–219.

INDEX